Palgrave Studies in the Future of Humanity and its Successors

Series Editors
Calvin Mercer
East Carolina University
Greenville, North Carolina, USA

Steve Fuller
University of Warwick
Coventry, United Kingdom

Humanity is at a crossroads in its history, precariously poised between mastery and extinction. The fast-developing array of human enhancement therapies and technologies (e.g., genetic engineering, information technology, regenerative medicine, robotics, and nanotechnology) are increasingly impacting our lives and our future. The most ardent advocates believe that some of these developments could permit humans to take control of their own evolution and alter human nature and the human condition in fundamental ways, perhaps to an extent that we arrive at the "posthuman", the "successor" of humanity. This series brings together research from a variety of fields to consider the economic, ethical, legal, political, psychological, religious, social, and other implications of cutting-edge science and technology. The series as a whole does not advocate any particular position on these matters. Rather, it provides a forum for experts to wrestle with the far-reaching implications of the enhancement technologies of our day. The time is ripe for forwarding this conversation among academics, public policy experts, and the general public. For more information on Palgrave Studies in the Future of Humanity and its Successors, please contact Phil Getz, Editor, Religion & Philosophy: phil.getz@palgrave-usa.com

More information about this series at
http://www.springer.com/series/14587

Alcibiades Malapi-Nelson

The Nature of the Machine and the Collapse of Cybernetics

A Transhumanist Lesson for Emerging Technologies

palgrave
macmillan

Alcibiades Malapi-Nelson
Toronto, Canada

Palgrave Studies in the Future of Humanity and its Successors
ISBN 978-3-319-54516-5   ISBN 978-3-319-54517-2 (eBook)
DOI 10.1007/978-3-319-54517-2

Library of Congress Control Number: 2017940038

Cover design by Emma Hardy

Printed on acid-free paper

This Palgrave Macmillan imprint is published by Springer Nature
The registered company is Springer International Publishing AG
The registered company address is: Gewerbestrasse 11, 6330 Cham, Switzerland

Contents

CHAPTER 1

Introduction

This book is concerned with the rise and fall of cybernetics, understood as an inquiry regarding the nature of a machine, and the lesson that transhumanist technologies can learn from that struggle.

Transhumanism, a growing global movement which aims at the amelioration of the human condition, can be understood as underpinning several contemporary projects in the so-called "emergent" (or convergent) technologies (biotechnology, information technology, nanotechnology, and cognitive science). Some of these projects, characteristically blurring traditional dichotomies (e.g., between the natural and the artificial, between a simulation and the simulated object), are usually instantiated in research on, among other fields, synthetic biology, Artificial Life, genetic manipulation, prosthetics, and cognitive enhancement. The general aim at "transcending" our traditional biological boundaries is clear, and its metaphysical overarching view is a radically mechanistic one.

Although explanatory mechanisms and mechanistic imagery are fundamental to modern science, a literal equalization (beyond a metaphorical identification) of natural phenomena with machinery is rather recent. This literal equalization was, in fact, the signature of "classical cybernetics". This 1940s movement, which experienced a relatively quick apotheosis and a rather abrupt collapse (it lasted for about a decade), is still subject of investigation, partly due to the growing awareness of the profound impact that it had in what eventually became current and familiar scientific

© The Author(s) 2017
A. Malapi-Nelson, *The Nature of the Machine and the Collapse of Cybernetics*, Palgrave Studies in the Future of Humanity and its Successors, DOI 10.1007/978-3-319-54517-2_1

outlooks. Transhumanism is not exempt from this shaping influence. Indeed, cybernetics could be understood as an instance of it. Given the driving force exerted by the transhumanist impetus on avant-garde technological undertakings, both their unifying thrust and its instantiation on various research and developments risk experiencing a fate similar to that of cybernetics.

Mechanical devices have certainly been the focus of attention in the past. This book is not concerned with those—interesting in their own right, but tangentially related to the notion of a machine. In fact, there seems to be so far no treatise which directly addressed the nature of a machine *per se*. Cybernetics is a relatively recent phenomenon in the history of science, and it is my thesis that it was fundamentally concerned with the notion of a machine. Furthermore, cybernetics might stand as the first time in history where the nature of a machine *tout court* was taken up to constitute the hinge upon which the whole enterprise revolved. This scientific movement, which has arguably not yet received adequate attention from a *philosophical* standpoint, will be addressed at fair length and depth, using it as a guiding instance for the questioning of the nature of a machine.

Delving deeper into the theoretical core of cybernetics, one could find that the contributions of William Ross Ashby and John von Neumann shed light onto the unique ways in which cybernetics understood the "essence" of a machine. However, both these contributions have been relatively relegated to a secondary role in historical treatments, leaving their philosophical import somewhat in the dark. Ross Ashby pondered about the useful outcomes of extending the scope of "the mechanical" after offering an account of the nature of a machine. This extension would later encompass realms that will show to be problematic for mechanization, and the way in which a machine-ontology was applied would trigger effects seemingly contrary to cybernetics' own distinctive features. Von Neumann, on the other hand, tinkered with a mechanical model of the brain, realizing grave limitations that prompted him to look for an alternative for cybernetics to soldier on. The proposal that came out of it might have signified a serious blow against cybernetics' theoretical core.

The contributions coming from both thinkers, in their own ways, spelled out the main tenets of the cybernetic proposal, consistent with their original intentions for further advancing it. But these contributions—likely against the will of the contributors themselves—might have led to cybernetics' own demise. The whole story can be framed under the rubric of a serious inquiry into the metaphysical underpinnings of a machine. The rise and fall of cybernetics could thus help us better understand what a machine is from a philosophical standpoint—and from there, provide transhumanist technologies with a solid ground to flourish.

This book partly aims at carefully unveiling the theoretical causes behind the implosion of the cybernetic enterprise. In fact, it advances the idea that the cybernetic understanding of the nature of a machine entailed ontological and epistemological consequences that created impossible conundrums (both material and theoretical) for the enterprise. These *ad intra* developments came from the cyberneticians themselves, out of an effort to further explore the deep mechanistic connections that they were unveiling. And these insights spelled doom for the enterprise. As a matter of conclusion, however, the book will propose that given both our current state of materials research and contemporary practices in scientific methodology, there might be indeed a way for cybernetics to bloom this time—instantiated in successful transhumanist (emergent) technologies. Since this development entails the biological and philosophical alteration of what we regard as "human", squarely addressing the possible ways in which this modification will take place seems to acquire a character of urgency.

The first two chapters (2 and 3) give a historical overview of cybernetics; they depict its origins, rise, climax and subsequent downfall. They aim to reveal how deeply intertwined the project was since its inception with an epistemology that directly stems from experimentation with machine behavior. Chapter 3 deals with the preceding events that lead to the emergence of cybernetics, mainly in recent history of mathematics. This chapter aims to show how a previous transformation of the notion of a machine paved the way for cybernetics to appear. Chapter 5's purpose is to philosophically dissect the main tenets of cybernetics, in order to unveil the metaphysical commitments behind the enterprise at large. This chapter shows how the entirety of the cybernetic ethos was underpinned by a framework constituted by bold conjectures regarding the nature of a machine. Chapters 6 and 7 address the contributions of Ross Ashby and John von Neumann, respectively. These contributions should help uncover relatively hidden tensions in the cybernetic project—both stemming from the nature of a machine. Chapter 8 ties up both Ashby's and von Neumann's contributions with the erosion of the main pillars of cybernetics. It shows how cybernetic inquiries into the nature of a machine might have proved fatal to the very fabric of the project, due to unsolvable theoretical tensions. The last two chapters anchors the cybernetic experience with the scope and limits of transhumanism. Chapter 9 establishes the historical and philosophical ancestry between cybernetics and transhumanism. It also unveils the unfinished (and somewhat surreptitious) transhumanist agenda underpinning much of current emergent technologies. The last chapter (10) proposes ways in which transhumanist emergent technologies can avoid the cybernetic fate that is otherwise likely bound to meet. Advances in materials research (e.g., nanotechnology), the vanishing of vitalistic preconceptions (i.e., biology as the study of natural machinery) and changes in scientific practice (i.e., the ontological status of simulations in scientific methodologies) can be the loci where the cybernetic impetus of transhumanism find fertile ground this time.

CHAPTER 2

Cybernetics: The Beginnings, the Founding Articles and the First Meetings

THE PROBLEM OF ANTI-AIRCRAFT WEAPONRY

On September 1, 1939, the Nazis invaded Poland. Two days later, both France and Great Britain declared war on Germany. Soon thereafter, constant aerial raids were launched from both sides. Anti-aircraft guns were proven reasonably effective against the massive air bombers that began to cast immense shadows over Europe. They flew at a relatively slow pace and in a fairly straight path. But the ground guns' effectiveness against the escorting fighter planes was wanting. These aircraft were smaller, faster and capable of snappily avoiding predictable paths with ease and surprise[1]—the celebrated and sometimes legendary skills of a pilot were largely based upon these extreme maneuverability capacities. Since a shot shell would take up to 20 seconds to reach the desirable altitude for explosion,[2] at a point 2 miles after the original position of the flying target,[3] a very concrete problem for the military emerged.

---

[1] Mindell 1995, p. 78.
[2] Conway and Siegelman 2005, p. 110.
[3] Mindell 2002, p. 86.

© The Author(s) 2017
A. Malapi-Nelson, *The Nature of the Machine and the Collapse of Cybernetics*, Palgrave Studies in the Future of Humanity and its Successors, DOI 10.1007/978-3-319-54517-2_2

At that point, inheriting the legacy of the knowledge gained from the first aerial strikes at the end of World War I, the human decision factor was being gradually merged with the relative reliability of a machine designed to fulfill a task. An early proto-computer (a "predictor") would receive visual information from soldiers on the ground, regarding the possible size of the target, its probable speed, the height of its path, the wind situation at the moment, and so on. The analog machine would then plot the direction toward which the soldier had to aim the gun. If all worked well, the soldier had to worry mainly about constantly and accurately pointing at the moving target and re-charging ammunition—while the predictor would take care of the angle of shooting. Ideally, this synergy of man and machine would create a cloud of exploding shells around the target's flying path. Shrapnel should sufficiently damage the aircraft's fuselage for it to go down, become an easier target due to damage, or abort its mission and retreat.

It is now widely agreed that the first air raids over the UK, the "Battle of Britain", was the first military failure for the German advance over Europe. Further, British retaliation against Germany proved to be devastating for the Nazis. The English air forces inflicted a serious blow to the Nazi leadership's image, coordinating bombings to happen precisely while important political speeches were taking place—topping it with one on Hitler's birthday. The German morale was however unbroken, and the retaliation against Britain, in terms of air strikes, was massive—it is referred to as the "Blitz" ("Lightning" in German). Despite the fact that the German mission was more successful this time, it still failed to accomplish its main goal: to cripple the military response capacity of the UK. In fact, raids over Germany, now with help from the allies, increased, and overall the destruction ended up being worse on the German side.

Although the USA chose to stay out of the war in the beginning, it gradually modified its stance, starting by helping its soon-to-be-allies with supplies—a position captured in President Franklin D. Roosevelt's famed "Arsenal of Democracy" catchphrase. American elites were, expectably, following the events closely. Despite an alleged majority opposing joining the war, sentiments of preparation and readiness for the dramatic possibility were strong. In fact, there already were secret talks taking place between Great Britain and America regarding possible courses of action in the eventuality that the latter would enter the conflict. Scientists and academicians were not excluded from the patriotic sentiment of becoming useful for their nation given the distinct possibility of war. Quite the contrary.

Norbert Wiener (1894–1964) was in his forties an already famed mathematics professor at the Massachusetts Institute of Technology (MIT). Of Russian Jewish ancestry, he married a bride brought from Germany, had two daughters, and ended up attending a Christian Unitarian church with his family. A child prodigy, Wiener already knew the alphabet at 2 years old, finished his mathematics bachelor's degree at Tufts University at 14, and after studying philosophy both at Cornell and Harvard universities, he ended up obtaining a Ph.D. in mathematics from Harvard University at 18. After further studying in Europe under Bertrand Russell and David Hilbert, he finally landed a position at MIT, where he remained for the rest of his career. Between his return from Europe and the MIT professorship, World War I was still being fought. Wiener attempted to enlist in the military, but he was rejected, twice—his eyesight was remarkably poor. The third time he applied, this time as a lowly infantry soldier, he was accepted. World War I ended, however, just days later and to his chagrin he was quickly sent back home.

World War II broke out while Wiener was teaching at MIT. This time, equipped with prestigious degrees and academic fame, he approached the military via a proxy more related to his skills, addressing the scientific bureaucratic body set up to face the war. Vannevar Bush (1890–1974), a doctor in electrical engineering also from MIT and President Roosevelt's consultant in all matters scientific, insisted on creating a special unit in charge of linking military and civilian knowledge. World War I allegedly showed that these two remarkably distinct and traditionally opposed human groups are not naturally inclined to work in tandem. Collaboration and synchronization between both had to be fostered—even enforced. Thus, the National Defense Research Committee (NDRC) was created, with Bush as its Chairman. Soon later, the NDRC, which had authority for organizing applied scientific research, but *not* for funding and developing projects, was superseded by the Office of Scientific Research and Development (OSRD). This newer office had the power to give the go-ahead to projects, enjoying access to practically unlimited resources for funding.

The NDRC, which ended up being the research branch within the bigger OSRD, was divided into five divisions:

Division A—Armor and Ordnance.
Division B—Bombs, Fuels, Gases, & Chemical Problems.
Division C—Communication and Transportation.
Division D—Detection, Controls, and Instruments.
Division E—Patents and Inventions..
Division D, the one that concerns this chapter, was itself divided into four sections:
Section 1—Detection.
Section 2—Controls.
Section 3—Instruments.
Section 4—Heat Radiation.[4]

Thus, the division and section to which Wiener had to apply was D-2, chaired by the doctor in engineering and mathematician Warren Weaver (1894–1978). In his letter, Wiener made sure that the upper echelon of the newly appointed scientific administration would acknowledge his eagerness to be at the service of his country through the military, clearly stating his wish: "I hope ... you can find some corner of activity in which I may be of use during the emergency".[5] Not long after, to Wiener's delight, he received word from NDRC letting him know that the military was interested in one of his proposals.[6]

---

[4] Mindell 2002, ch. 7.

[5] Galison 1994, p. 228.

[6] Indeed Wiener proposed several projects. Among them, he was advocating for the positioning of immense aerial balloons to be massively ignited in a prolonged manner once the enemy would fly close-by. This rejected proposal might have been refused due to its "passive defense" character. There already were projects that came to fruition which placed static "barrage balloons" tethered to the ground. This would provoke two desired outcomes. The enemy planes would have to fly higher, thus becoming more available as targets—which is more difficult when they fly at a low altitude. Also, if they would refuse to fly higher, they would risk becoming entangled with the wires and crash. As it turned out, pilots soon easily overcame this difficulty, and these defenses stopped being used (Galison 1994, pp. 228–229).

The UK had secretly developed a "cavity magnetron"—a type of radar far more advanced than anything similar developed anywhere else. This was probably Britain's best kept "secret" at the time (the Germans eventually found out). The British were willing to share this ground-breaking weapon accessory with the Americans in exchange for help in quickly developing anti-aircraft weapon systems—with which such an invention should interface. Vannevar Bush was aware that American anti-aircraft technology was not precisely cutting edge, having barely improved after World War I. Closely following the manner in which the war was unfolding in Europe, he understood that such was precisely the kind of weapon improvement that America urgently needed in case the war reached the Pacific coast. He agreed to the exchange. America would be helping the UK against the advancement of the "Axis" (Germany, Italy, and Japan) while dramatically improving its own anti-aircraft defense.

Early on, Wiener took issue with a perceived inadequacy of contemporary anti-aircraft weapon systems, thus wishing to approach the problem from a different angle—from a perspective closer to his own fields of expertise. In the past, Wiener had spent years working on the mathematical tractability of diverse chaotic phenomena, trying to find recognizable patterns via formulaic equations. For instance, he devised a numerical "filter" for the allegedly random movement of dust particles (Brownian motion)—which was later known as the "Wiener filter". Given the relative ineffectiveness of the current anti-aircraft predictors, he saw a probable application of similar principles for the construction of an improved machine—one that could predict the position of the flying target by extrapolating its likely future location, statistically deducing it from its previous locations. Taking into consideration all the relevant variables available and finding an "average", the machine would predict where the enemy object would be in the near future, thus producing its best possible shot at neutralizing it.

THE AA-PREDICTOR

The utmost importance of interfacing pure theory and embedded practice was always in Wiener's mind—and in a personally dramatic manner. His plans for earning a Ph.D. in biology were trumped by his utter manual clumsiness while making his way through the mandatory laboratory coursework.[7] Once a

---

[7] Hayles 1999, p. 99.

professor, his poor sight led him to famously walk down the campus hallways with one hand always touching the wall, allegedly following the belief that tracking a maze's wall would eventually take him to the exit. Wiener's inability to shoot and hit even a big immobile object, "a barn among a flock of barns",[8] spelled for him, much to his dismay, non-admittance to military service. The reason why he famously walked with his head tilted back was that he instructed his optometrist to divide his thick glasses into two: the upper half would allow him to read, whereas the lower half to see afar. Thus, he used to walk toward and approach people in a manner of looking down on them—which gained for him some antipathy. He was just severely myopic.[9]

It is then unsurprising that as soon as he received funding for his project, finding a person with the appropriate hands-on skills was the first task. Nobody without at least passing knowledge in applied electronics (e.g. radio building) would even be considered as a potential assistant. "There is nothing in a drawing in abstract algebra or topology...which would prepare one in a way to cooperate with engineering design".[10] From now on, the scientist involved in wartime projects would have to get out from the comfort zone of "clean" and pure science and "get dirty", so to speak, with real-world constraints.

> If I could not find these talents joined together in a single man, I would be forced to assemble a team of people each with particular talents in one field and a general knowledge of the others. In this team, I would probably be the only mathematician, thus, the project as a whole would concern other engineering groups as much as the American Mathematical Society and it would be necessary for me to cross ordinary professional lines.[11]

In January 1941, Wiener hired Julian Bigelow, a talented and down to earth electrical engineer from MIT. Most relevantly, he was an active airplane pilot as well. He joined Wiener's team for exclusively working in

---

[8] Conway and Siegelman 2005, p. 124.

[9] For the famous anecdotes about his legendary blindness and eccentricities, see Jackson 1972.

[10] Galison 1994, p. 235.

[11] Ibid.

the circuit design of the predictor, but his skills would later prove to be essential for a much bigger project.[12] As alluded to above, Wiener wished to utilize knowledge gained from his previous mathematical investigations into chaotic motion in order to improve the capabilities of the current predictors—so that the machine could better foretell the future position of a flying enemy target. But that part was just half the story. Such an approach, even if potentially of great value, had to be connected with the unavoidable "real-world" constraints in order to bring something truly useful—for the military at any rate—to fruition. Fortunately, Wiener's knowledge proved to be capable of modeling and articulating Bigelow's grounded know-how expertise as a pilot in interesting ways.

Bigelow emphasized to Wiener that when a pilot is shot at, the scope of possible sudden movements he would try in order to avoid enemy fire was not boundless. The dynamics of contact between the vessel and air, the resistance of the latter against the former's fuselage, the torsion forces that the airplane's body can maximally handle... all those aspects form the physical landscape that limit the amount of actual "moves" and "tricks" available to the pilot. Indeed, a wrong sudden movement could end up breaking the plane. Wiener grasped the nature of this limited range of actions crisply. He then stated that there was a way to insert this limitation into the operation of the machine. Statistics was to provide the way: since the limited range of movement was known, one could find an average for the likelihood of a certain move to occur. In fact, the same reasoning could also be applied to the infamous sluggishness of the heavy anti-air guns operating from the ground. The predictor to be built was going to dramatically improve its prognostications using the power of statistics. In Wiener's mind, that was the feature that was going to provide this machine its unique capabilities.

As mentioned above, toward the end of World War I interest in anti-aircraft weaponry substantially grew. Accordingly, the notion of a "servo-mechanism" was given a good deal of attention, and thus it was right from the start at the crux of Wiener's designs. Indeed, the patent for which he applied was firmly within the realm of predictors—themselves paradigmatic

---

[12] Bigelow would go on later to fulfill an important role as an irreplaceable collaborator for similarly inspired projects that came after this first one, which spanned a new way of thinking about science, engineering, and even philosophy, which Wiener later baptized as "cybernetics" (more below).

examples routinely articulated in treatises about servomechanisms. Although most work had been done "empirically"—without a "school of thought" or anything pertaining to a mature technology—there were two main writings circulating around the American war laboratories. One was by Hendrik Bode—a doctor in engineering working from Bell Laboratories— and another one, less exhaustive but more encompassing, by LeRoy MacColl—a mathematician from Columbia University. Wiener chose MacColl's treatise[13] and soon introduced it to Bigelow.

The common root between "servomechanism" and "servant" is not coincidental. Commonly regarded as an Anglicism, the noun "servant" was adopted in its unchanged form from the French language. The French *servant* denotes an entity fulfilling a role, but it has the connotation of being hired—and thus, remunerated. However, the older Latin *servus* (from where the French word itself derives) has a more direct referent to the notion of a "slave". A *servomotoeur* (French) is an engine subservient to a greater, more encompassing mechanical task. Its enslavement is firmly put in place by means of a phenomenon within control theory whose rediscovery was in itself, for Wiener, a sort of revelation: negative feedback. This was a type of corrective input that would fix possible deviations from the system's designed task. As we will see later, negative feedback as a major useful element entered Wiener's mind after facing the very real-world problems he was choosing to confront. Of the two classical writings on servos alluded to above, the one chosen by Wiener (MacColl's) treated servomechanisms as paradigmatically "enslaved" by means of negative feedback. Thermostats and self-guided torpedoes would be prime examples of devices reliant on these servomechanisms. In fact, Wiener would not regard his own system in a fundamentally different manner than these two devices. He theoretically expanded the main tenets underlying one self-regulated system (that of a thermostat), to envision another similar one (that of an anti-aircraft weapon system). Seen from afar, both systems, thermostat and anti-aircraft weapon system, would be one and the same.

The attentive reader will notice that while in the thermostat all the comprising parts of the closed system are inanimate and mechanical in nature (gears, levers, etc.), Wiener's system will have human beings at two points (the pilot and the gunner) interlaced with plane and gun. It did not take long for Wiener to see, after he began working out the systems' circuits, that those two

---

[13] MacColl 1945.

entities would be the source of irregularities for the predictor. "Irregularities", however, were not a scary notion for Wiener. He had been dealing with phenomena of this sort while studying the allegedly random (Brownian) motion of dust—hence the confidence in his own usefulness for the cause of the war. Thus, Wiener proceeded to treat them mathematically. This mathematization would pre-require some sort of measured report of these irregularities, captured within a controlled environment.

As advanced above, Bigelow's input would prove itself essential for the construction and operation of the new predictor—and more. As his airplane piloting know-how was in fact complementing his remarkable engineering skills, he designed and built a secondary machine at Wiener's laboratory that would begin producing the data needed for the principal machine, the predictor. Bigelow's machine, once constructed, begun to receive test subjects in a laboratory-controlled environment, obtaining valuable data from their behavior.

The "gunners" recruited to use Bigelow's machine had to align, via the manipulation of a delayed-effect joystick, two large light dots projected against a wall—a white one representing the enemy plane and a red one being the gunner's beam. The recreated situation aimed to simulate the experience of a gunner aiming his heavy and slow gun turret toward the flying target and successfully hitting it with exploding shells. As mentioned before, airplane fighting made of the avoidance skills a matter of celebrative tradition—after all, the survival of the pilot was the positive outcome. This skilled irregular flying pattern had to be somehow represented in the movement of the white beam. Accordingly, Bigelow had the ingenious idea of projecting the white dot on the four walls of the classroom in a circular manner. When the dot would pass from one side of the corner to the other, it "jumped", mimicking the sudden and deceptive avoidance behavior of a trained pilot. The "gunmen" reported how difficult this sudden move made the task of aligning their beam to the target, having to move the rigged joystick well ahead before the white spot would arrive at the predicted target location. Wiener was happy with the machine's simulation, and the data being produced by the gunners' behavior was later input into Wiener's predictor.[14]

---

[14] Peter Galison has a different view on the nature of Bigelow's machine, which allegedly recreates the kinaesthetic dissociation present in a pilot's manoeuvres. This is how Wiener describes the psychological phenomenon: "the pilot's

Around the same time, and unbeknownst to Wiener, Warren Weaver was (due to war urgency) funding another anti-aircraft weapons system project via NDRC's Fire Control section—of which he was chairman. Of special priority was the quick construction of a predictor that could be interfaced with Britain's shared radar machine. At the time, up to 14 people were required to provide the visual input and manual direction of the heavy and slow gun turrets. In Wiener's mind, this process was "fascinating, absurdly cumbersome and ripe for improvement".[15] Given the relentless German bombing over England, automation of the whole system indeed carried a sense of urgency.

Neither Wiener nor Bigelow, despite their fairly high security clearances, knew at the time of the Radiation Laboratory[16] functioning in a close-by building at MIT itself. The British secret invention was being studied there. Both were invited, however, to the Bell Laboratories, where a simpler predictor, soon to be coupled with the British radar, was being constructed under the supervision of that other theorist of servomechanisms, Hendrik Bode. The engineers at Bell were dismissive of Wiener's

kinaesthetic reaction to the motion of the plane is quite different from that which his other senses would normally lead him to expect, so that for precision flying, he must disassociate his kinaesthetic from his visual sense." (Galison 1994, p. 236). The air resistance against the plane's steering parts provides a feedback to which the pilot's own visual input has to eventually fit. It might give the sensation that the machine itself is resisting its direct control, and learning how to dominate it requires skill and experience. In fact, the tradition of giving proper names to flying machines might actually stem from this "wrestling" between man (what the eye provides) and machine (what the physical controls give back in response). For Galison, Bigelow's machine seemingly represented not the enemy pilot (white dot) and the friendly gun-man (red dot), but the enemy pilot (white dot) and the friendly pilot (red dot). Galison used a vague recounting from a later book of Wiener (1956, pp. 250–255), and a report by the American mathematician George Stibitz. Conway and Siegelman used declassified records from 1941, where Bigelow himself explains the machine workings at a Bell conference. The amount of quotes of Bigelow himself seems to make Conway's understanding the right one. See Conway and Siegelman 2005, pp. 110–115.

[15] Conway and Siegelman 2005, p. 110.

[16] Rad Lab, in short. This nickname was a *non sequitur* from the original one, in order to trigger confusion and maintain its secrecy (Conway and Siegelman 2005, p. 115).

project, feeling that theirs was less complicated and still able to accomplish the same task. Wiener and Bigelow were unmoved, pointing out the almost simplistic and unsophisticated nature of the machine built by these engineers. They did not take the statistical considerations pointed out by Wiener for its construction, and it was exclusively based on calculating the future angle of the plane using its previous location. When the predictor was coupled to an anti-aircraft gun and tested in the field, it did fairly well—but far from perfectly. However, it did not have to be particularly accurate, since the shells were altitude-sensitive, and they would likely explode close to the plane fuselage anyway, damaging it with shrapnel. Given the urgency of the situation, Wiener and Bigelow gave a cold approval of the project, realizing that the Bell predictor had to be deployed anyway given the Axis' ongoing devastation.

Not long after Wiener was finally notified of the secretive Rad Lab operating very close to his own, and was asked to train the group of young scientists gathered in there. Since Wiener's ultimate reason for working on this whole project was patriotic in nature, he put understandable professional rivalries aside and agreed to help. However, schooling young and eager Ivy League minds would not be an easy task. From the point of view of scientific practice *vis à vis* different disciplines, there was a feature that lay at the center of Wiener's operational concerns. If we follow a traditional distinction between the notions of multidisciplinarity, interdisciplinarity, and transdisciplinarity,[17] one could arguably make the point that the scientists recruited by the bureaucratic overseeing institution (NDRC) would likely come equipped with a multidisciplinary perspective—open to what other disciplines would have to offer, but ultimately unwilling to substantially cross their own field boundaries. Wiener expected for his project (and indeed for any *working* project) in contrast, a radical interdisciplinarity—based upon what other fields could provide, holistically creating a separate entity. Partly due to adherence to already familiar ways and reluctance to change (a usual signature of large administrative bodies), this pre-requirement proved to be particularly

---

[17] "Multidisciplinarity draws on knowledge from different disciplines but stays within their boundaries. Interdisciplinarity analyzes, synthesizes and harmonizes links between disciplines into a coordinated and coherent whole. Transdisciplinarity integrates the natural, social and health sciences in a humanities context, and transcends their traditional boundaries." (Choi 2006, p. 351).

difficult to fulfill—so difficult in fact that the project was at some points almost brought to a permanent halt. Wiener complained . . .

> New members of the staff of your Laboratory are recruited from the theo-
> retical physicists or mathematicians of the country, or indeed anywhere
> except from among the ranks of communication engineers in the strictest
> and narrowest sense of the term . . . to turn such an individual loose in your
> laboratory without special training, no matter what a big shot he may be in
> his own subject, is like ordering a corn-doctor to amputate a leg.[18]

The anchoring in reality by means of machine-construction should provide a hub for the encounter of heretofore separated disciplines. They all had to converge in one physically embodied project. This was an instance where an applied interdisciplinarity had to be the norm.[19] Realizing that it was not going to happen, Wiener renounced this project. With renewed motivation, the scientific duo returned to its own project, aware that theirs had the potential to engender a much better prognosticator. A considerable amount of data was produced by Bigelow's machine, and the results finally began to be entered and crunched by Wiener's new predictor.[20] Soon the mathematician

---

[18] Galison 1994, pp. 240–241.

[19] As we will see later, a similarly inspired transdisciplinarity—the encounter of several disciplines, natural and social, that would as a whole transcend boundaries—was about to occur just a few years down the road under the banner of "cybernetics".

[20] The machine would be fed with data produced by observation of aircraft behavior. Beforehand, Wiener's statistics would predict where the target would be 1–2 seconds before it reaches its position, extrapolating from observations taken 10–20 seconds in the past. All this information would be input into the machine, which would update itself every time new "real-life" input is fed into it, rendering in advance the precise spot where and when the gun should fire. Since Wiener's machine compressed time by a factor of 4–5, a 1-second prediction would amount to 10 seconds in real life. And since a shell would take 20 seconds to explode from the moment it is shot, it would seem that Wiener's machine was indeed in the right track. Wiener's lay man explanation of his machine goes as follows:

> The proposed project is the design of a lead or prediction apparatus in
> which, when one member follows the actual track of an airplane, another

grasped two identifiable phenomena. One seemed positive for the project, the other one less so.

On the one hand, pilots were exceedingly consistent while in the simulator. This finding seemed to match Wiener's assumption that humans under stress tend to act repetitively. For Wiener, this also cast a light on the apparatus, which made it look like a precursor to a learning machine. The predictor seemed to be good at learning how to accomplish a pilot's death in a rather "personalized" manner. Speculating, the machine could be successful at knowing how to effectively kill *one* individual, statistically extrapolating data based upon his past behavior. In the area where the predictor was shining, it seemed that it was indeed managing to foretell the future. As the

> member anticipates where the airplane is to be after a fixed lapse of time. This is done by a linear network into which information is put by the entire past motion of the airplane and which generates a correction term indicating the amount that the airplane is going to be away from its present position when a shell arrives in its neighborhood (Masani 1990, p. 182).

David Mindell explains it with somewhat more detail:

> Wiener and Bigelow...turned to statistics and designed a new predictor based on "a statistical analysis of the correlation between the past performance of a function of time and its present and future performance." The network calculated a future position of the target based on the statistical characteristics of its past performance (its autocorrelation). It then continually updated its own prediction as time passed, comparing the target's flight path with previous guesses. A feedback network converged on guesses that minimized this error (Mindell 2002, p. 278)

For a description of Wiener's statistical prediction, see Galison 1994, p. 237. For an accessible explanation of the machine's inner workings, see Mindell 2002, ch. 11. For a technical account of Wiener's machine, see Bennett 1993, ch. 14.

Given the later fate of this machine, the reader should bear in mind that its importance for cybernetics is not anchored in the details of its inner workings. Instead, its relevance is based upon the insights it provided Wiener while trying to "mechanize" the behavior of both the pilot and gun-man, making it overlap with that of a machine.

visiting supervisor George Stibitz put it in his working diary, referring to the impression the machine caused on Warren Weaver...

> It simply must be agreed that, taking into account the character of the input data, their statistical predictor accomplishes miracles... For a 1-second lead the behavior of their instrument is positively uncanny. WW [Warren Weaver] threatens to bring along a hack saw on the next visit and cut through the legs of the table to see if they do not have some hidden wires somewhere.[21]

Indeed, some technicians jokingly made allusions to the "gremlins" that were surely inhabiting and controlling the machine—venting the feeling of "uncanniness" present when the non-living behaves in a life-like manner. Warren Weaver, after acknowledging the important result, pointed out that now the challenge remained regarding inter-facing it with the rest of the weapon's system in order to deploy it in the field. After all, it still had to be hooked to the British radar invention and the complicated manned gun turret. This proved to have many real-world details that presented concrete problems for Wiener and Bigelow.

Coupling machines in order to accomplish a task was of course not a new feat. In fact, history registers several accomplishments where inter-regulated interface mechanisms worked successfully—from the Ancient Greeks on.[22] However, there were some instances where they seemed to not work in tandem—and for somewhat mysterious reasons. For example, gun turrets placed in European naval vessels during the eighteenth century sometimes tended to engage in recurrent "spasms", swinging erratically and out of control for periods of time.[23] A similar occurrence was later recognized in sound engineering. A microphone and a speaker in close range would trigger a high sound loop—the so-called "howling" phenomenon—when the amplification of both sources went unlimited.[24] As Bigelow reminded them, these instances of overcompensation in machine

---

[21] Galison 1994, p. 243.
[22] See Mayr 1970, ch. II.
[23] See Bissell 2009.
[24] Ibid.

interfaces were gradually fixed by "negative feedback": capturing a portion of the output and feeding it back into the stream, balancing it.[25] Wiener was heavily impressed by the importance of this negative feedback in mechanisms with a purpose or task built-in. He weighed the revolutionary importance that this engineering insight had for the Industrial Revolution[26] and did not hesitate in asking help from an old friend, who allegedly found similar occurrences in living organisms.

Arturo Rosenblueth, a Mexican Jew of Hungarian heritage who studied medicine in Paris and Berlin, was the protégé of Walter Cannon—a famed biologist who pioneered work with self-regulating mechanisms in living tissue (homeostasis). Wiener met Rosenblueth a decade before at the Harvard Medical School informal dinners organized by the latter. These dinners were an environment for free intellectual discussion on scientific matters. Both men being avid talkers who liked to convey their ideas and

---

[25] If the output of a system suffers variations that preclude the system from completing its task (due to changes in its input), a portion of the output could be fed back into the input, thus balancing its output. In Wiener's words, negative feedback can fix the system's "defective behavior and its breaking into oscillation when it is mishandled or overloaded". It is "negative" because "the information fed back to the control center tends to oppose the departure of the controlled from the controlling quantity" (Wiener 1948a, pp. 97–98). Ashby uses the following explicative example:

> A specially simple and well-known case occurs when the system consists of parts between which there is feedback, and when this has the very simple form of a single loop. A simple test for stability (from a state of equilibrium assumed) is to consider the sequence of changes that follow a small displacement, as it travels round the loop. If the displacement ultimately arrives back at its place of origin with size and sign so that, when added algebraically to the initial displacement, the initial displacement is diminished, that is, is (commonly) stable. The feedback, in this case, is said to be "negative" (for it causes an eventual *subtraction* from the initial displacement) (Ashby 1956, p. 80).

[26] Embodied in the important role of James Watson's mechanical "governor", later mathematically treated by James C. Maxwell (1868). More on this in Chapter 3, Section "Norbert Wiener's *Cybernetics*". For a brief history of control servomechanisms in the last century, see Bennett 1996.

research with passion, Wiener clicked with Rosenblueth at once, and kept uninterrupted contact after that.

Upon being told of the problem Wiener was dealing with, Rosenblueth found it remarkably similar to a reported phenomenon occurring in some human patients: "purpose tremor". This situation would manifest in the person when she would like to perform a simple, isolated, modular, task—like picking up an object from a table. The person's arm would instead engage in uncontrollable swinging, back and forth, missing the object. If the person were handed a piece of candy, and was instructed to put it in her mouth, a similar occurrence would happen, missing the target—in this case, her mouth. Rosenblueth offered his insights to Bigelow and Wiener, and at least in theory, they seemed to have helped them find a solution. However, the implementation of these new self-regulatory aspects in the actual, physical instances of the machinery would prove to be a big challenge. Nevertheless, they soldiered on—as the devastating on going war demanded.

There was another realm that emerged as fundamental for accomplishing the task of building the new predictor. Although this aspect at first looked like a problem just for Wiener's project, it planted the seed of a far-reaching change of outlook. In fact, it expanded Wiener's framework to areas never before successfully treated mechanically. It was said above that Wiener found the pattern of each pilot's behavior consistent with itself on an individual basis. However, the distressed behavior from one pilot to another did not seem to reveal any common pattern. The machine was falling short at capturing a recognizable behavior pertaining to humans *in general*—in order to wipe them out effectively, which was the main point of the military-funded project in the first place. The human element in the system (the gunner and the pilot) was, as previously said, the main source of irregularities. Thus, the issue of "behavior" in general suddenly occupied the center of Wiener's concerns. Given his general background in undergraduate and graduate studies in philosophy and biology, Wiener was well aware of the school of "behaviorism" in psychology. Considering his own struggles regarding the rigorous articulation of behavior in general, he referred to the contemporary scholarship on the subject as having missed the point:

> Behaviorism as we all know is an established method of biological and psychological study but I have nowhere seen an adequate attempt to analyze the intrinsic possibilities of types of behavior. This has become necessary to

me in connection with the design of apparatus to accomplish specific purposes in the way of the repetition and modification of time patterns.[27]

To be sure, preoccupation with human behavior as a source of randomness and unpredictability—potentially fatal when present in the chain of command—was naturally at the core of the military mind. In the line of attack or defense, the human factor was sometimes regarded as less reliable than the mechanized parts of the weapon system—itself less prone to break, more predictable, and hence more reliable. As a veteran Admiral put it to a group of NDRC scientists when they were invited aboard a war naval vessel, "Twenty-five hundred officers and men: gentlemen, twenty-five hundred sources of error."[28]

Although the problem of (human) behavior was part and parcel of the nature of military command, now Wiener had this problem affecting the very design, construction and probability of success for his predicting machine. Wiener realized that the notion of *behavior* was at the crux of both his empirical and theoretical dilemma. Given the state of scholarship in psychology at the time regarding this issue, where "no behaviorist ha[d] ever really understood the possibilities of behavior",[29] and considering the urgency inherent to war-related enterprises, Wiener exercised a philosophical extension that would later prove to be prescient.

It does not seem even remotely possible to eliminate the human element as far as it shows itself in enemy behavior. Therefore, in order to obtain as complete a mathematical treatment as possible of the over-all control problem, it is necessary to assimilate the different parts of the system to a single basis, either human or mechanical. Since our understanding of the mechanical elements of gun pointing appeared to us to be far ahead of our psychological understanding, we chose to try to find a mechanical analogue of the gun pointer and the airplane pilot.[30]

[27] As expressed in a letter to his friend the British geneticist J. B. S. Haldane (Kay 2000, p. 80).

[28] Edwards 1996, p. 206

[29] Galison 1994, p. 242.

[30] Wiener 1956, pp. 251–252.

Indeed, given the state of darkness regarding the study of behavior in general and human behavior in particular, extending the realm of the mechanizable seemed like the logical course of action. Surely much needed to be understood regarding the new realm of control theory. But it struck Wiener as evident that such knowledge, even if limited, was still greater than knowledge about man. Thus the source of irregularities (namely, the human element in the system) could be identified with a tractable aspect within the realm of machinery...

> We realized that the "randomness" or irregularity of an airplane's path is introduced by the pilot; that in attempting to force his dynamic craft to execute a useful manoeuver, such as straight-line flight or 180 degree turn, the pilot behaves like a servo-mechanism, attempting to overcome the intrinsic lag due to the dynamics of his plane as a physical system, in response to a stimulus which increases in intensity with the degree to which he has failed to accomplish his task.[31]

Identifying that the pilot's behavior is comparable with (as in "is amenable to be subsumed under the study of") servomechanisms, expanded the power of what could be mechanically tractable. From that point on, the usage of the term *behavior* (and its derivations) by Wiener and colleagues went well beyond the usual understanding of the notion as an exclusively Skinnerian concept—which also referred to a human being in machine terms, after equating it with a black box. However, as George Stibitz wrote in his report, Wiener's ideas, even if enriched, were still unmistakably anchored in clear behavioristic roots. He acutely pointed out that Wiener's "extension" of what behaves machine-like was based not upon the structure of the entity, but upon its behavior.[32]

> W[iener] points out that their equipment is probably one of the closest mechanical approaches ever made to physiological behavior. Parenthetically, the Wiener predictor is based on good behavioristic ideas, since it tries to

---

[31] Galison 1994, p. 236.

[32] This dichotomy will eventually be crucial for the enterprise that was about to emerge—cybernetics (more of this on Chapter 5, where its philosophical backbone is articulated).

predict the future actions of an organism not by studying the structure of the organism but by studying the past behavior of the organism.[33]

Regardless of the project's outcome, Wiener saw a new and powerful paradigm beginning to take shape, capable of encompassing realms heretofore banned from a purely mechanical discourse. He did not hide his enthusiasm toward what was emerging before his eyes.

At this point, the machine was tested only in the laboratory and with data obtained *ad intra*. Let us recall that so far, the data was produced by Bigelow's simulator in the carefully measured environment of the small MIT laboratory assigned to Wiener. Before the predictor could be deployed to a field test (like the one Wiener and Bigelow attended for testing the Bell predictor), they needed to input real-world data. For this, both men toured several proving ground facilities, getting all the information they could lay their hand on regarding aircraft behavior. A particular piece of data was especially useful. Flights 303 and 304 from the North Carolina proving ground registered telemetry that reported aircraft behavior every second. This was exactly what Wiener and Bigelow needed. Equipped with it, they crunched the data with the predictor for the next 5 months.

In practical terms, the results were not too encouraging. Wiener's predictor was compared with the mentioned simpler—and tested in front of him—machine constructed under Bell's Hendrick Bode. Bode's predictor relied solely on a calculation of the target's location based upon input angles and distances, and a re-computation every 10 seconds to correct the aiming. It did not have statistical analysis, unlike Wiener's, but it was based upon readily available technology—and hence it was comparatively easier to construct and deploy. Wiener's machine, with all its bells and whistles, performed barely better than Bode's for flight 303 and even slightly worse for 304. The attack on Pearl Harbor had just occurred and imminent pragmatic results were urgently needed: The USA had finally entered the global conflict. Wiener and Bigelow realized that hundreds of telemetry samples would need to be input into the machine in order to make it practically functional, and the chances of having it finished before the end of the war were slim. They had to agree that

---

[33] Galison 1994, p. 243.

Bode's machine was the one to be favored by the military, and thus abstained from recommending their own machine for construction.

Given his relentless patriotic spirit, Wiener was obviously disappointed in failing to provide tangible help by means of this project. However, he did substantially collaborate, in the larger scheme of things, to the cause of the war. Wiener was acknowledged—by both the engineers behind Bode's predictor and the scientists at the Rad Lab—as responsible for improving the machine that the latter had already constructed. He would even go on to help "on the field" when this "rival" predictor would malfunction (e.g. aboard a naval vessel).[34] Furthermore, he produced by this time (1942) a manuscript summarizing the evolution of the theory of servomechanisms up to that point (improved with his insights on negative feedback), entitled "The Extrapolation, Interpolation and Smoothing of Stationary Time Series: With Engineering Applications". The piece was immediately labelled as "classified" the moment Warren Weaver received it.[35] The "Yellow Peril"—as it was nicknamed due to both its staggering complexity and the yellow cover proper to classified documents at the time—became a legend of sorts among military engineering circles.[36]

The apparent failure by Wiener to accomplish the task of a new, substantially enhanced predictor yet opened up new routes of thinking. In the attempt at constructing this ambitious predictor, Wiener had had a profound insight—one that laid the foundations of a new science that would soon be regarded as revolutionary on a global scale.

THE PRE-MEETINGS AND FOUNDING ARTICLES

Before the predictor project came to its conclusion, Wiener already had grand plans stemming out of his experience. Indeed, it was reported that at the time he was working on the machine, his mind was seemingly already on something else—as reported by his direct project supervisor, Warren Weaver. One should also take into account Wiener's outspoken

---

[34] Conway and Siegelman 2005, p. 124.

[35] Later declassified and published under the same title (Wiener 1949).

[36] Conway and Siegelman 2005, pp. 116ss. Wiener's contribution to the war effort and science in general was later clearly acknowledged by his government. He received the National Medal of Science from the hands of President Lyndon B. Johnson in 1964.

dislike for war-related secrecy—parameters which, to his understanding, run against the very notion of scientific practice. Wiener needed a freer environment to develop his new insights.

The Josiah Macy Jr. Foundation is a funding institution rooted in a wealthy family of merchants, interested in the interdisciplinary advancement of sciences—particularly medical sciences.[37] In 1942 it organized a gathering of 20 scientists to take place in New York, on May 13 and 14. Wiener seized the opportunity.

The conference was supposed to deal with issues of hypnosis and conditioned reflexes. In its spirit of interdisciplinarity, related approaches were welcome, and so Wiener's findings would allegedly fit. Wiener himself could not attend: the deadline for presenting his predictor was coming close. The person in charge of conveying his newly found ideas was a good choice. Arturo Rosenblueth, the medical scientist referred to above, who helped Wiener theoretically dissect the disconcerting endless loops, was to channel these ideas to the world for the first time—although to a small group, to begin with. This aspect was in fact the remarkable feature of the presentation, since much of the punch of Wiener's novelty had precisely to do with these circular loops disentangled with Rosenblueth's help. And indeed, Rosenblueth did deliver. Although no written records remain of this "Cerebral Inhibition Meeting", personal recounts of the gathering described it as intense and powerful.[38]

Rosenblueth carefully conveyed what was possible without breaking the war-related oath of secrecy—for example, he left aside much of the technicalities of Wiener's "Yellow Peril". He managed to deliver the core notions that the three men dealt with during work on the predictor. Rosenblueth reported that there was a novel notion of causality in nature that should be understood as "circular". More specifically, given that some sub-parts of relatively complex systems sometimes seem to be caught in endless loops, there seem to be cases where overcompensating forces push the system "out of whack" (e.g., wild swinging of a person's arm when she wants to take a bite of food, with tremor illness, missing her own mouth, or an anti-aircraft cannon wildly oscillating back and forth without

---

[37] It is still operating. See macyfoundation.org

[38] The invited anthropologist Margaret Mead reportedly broke her tooth in this meeting, and she did not realize it until it was over, due to the sheer excitement. See Heims 1991, pp. 14–17.

apparent cause). However, there is a countervailing input, a "negative feedback", that seems to correct this potential error, aligning the system back on track. A typical example would be that of a torpedo guiding itself, making the necessary corrections, so that it keeps aiming at its target (a ship or submarine) by following the noise of its machinery or the magnetic force exerted by its hull. Both the homeostatic self-regulation of body temperature in animals, and the phototropic behavior of a plant, are no different in substance. Rosenblueth, thus, stated that approaching these phenomena without understanding these occurrences within the framework of a "circular causality" would miss the mark for appropriately understanding the array of causes at play. This remark raised a few eyebrows.

Among the attendees were the neurophysiologists Warren McCulloch and Rafael Lorente de Nó, the neurologist and psychoanalyst Lawrence Kubie, and the anthropologists Gregory Bateson and Margaret Mead—a married couple. What Rosenblueth was saying felt for some of them like an escape valve out of years of frustration in their research and the scholarly literature of their own fields. Lawrence Kubie turned from neurology to psychoanalysis precisely due to what he would regard as intolerable reductions of the human psyche. The Spanish-American neuroscientist Lorente de Nó had found evidence for continuous loops in neural activity some years before—but no paradigm in which he could make sense of them. Bateson and Mead decried the lack of a sufficiently rich model to understand cultural exchanges and developments. In the case of Warren McCulloch, it could even be said that he saw light after years of darkness. McCulloch, as Lorente de Nó, found that neural activity regularly engaged in these aforementioned endless loops—without being able to make heads and tails of them either. With one aggravating detail. For decades McCulloch was trying to build up a model that would correlate the all-or-nothing character or neuronal firing with the all-or-nothing character of logical processes—trying to establish a hylomorphism between a neuron and a logical gateway. These "loops" were a major obstacle against this otherwise sophisticated mapping. And now there was a glimmer of light at the end of the tunnel.[39]

The organizer, Frank Fremont-Smith, happy to see this positive response, asked the foundation for a follow up—probably as a regular

---

[39] I will come back to this later in this chapter.

recurring gathering. It turns out that Warren McCulloch was already lobbying for it. Nevertheless, the pressures of the war, in terms of both contracted commitments and imposed secrecy, made everyone agree that it should be postponed until it was over. However, the wheels for something much greater in scope at many levels were already in motion. The next year, Wiener, Rosenblueth, and Bigelow published an iconic paper in the journal *Philosophy of Science*, largely based upon the presentation delivered by Rosenblueth, entitled "Behavior, Purpose and Teleology".[40] There Wiener and company went the extra mile into controversy. After reconfirming what was said in the meeting, they advocated for the need to revive the notion of a final cause, which allegedly is scientifically perceived in nature's doings. Indeed, they proposed the appropriation of the ages old notion of "teleology" in order to understand this reality scientifically. The proposal, available this time to a wider audience, raised many more eyebrows, but now the effects were polarizing.

In Aristotle's theory of causality, of the four causes that explain the being of things (material, formal, efficient and final), the final cause (*telos*) was the most important for understanding things in the universe. His whole ethics stems from the notion of telos applied to the thing "human" (*eudaimonia*). This *telos*, after having received a profound theological connotation during the Middle Ages, was later rejected by the spirit of the Scientific Revolution—relying now on Bacon's *experimentum crucis* as the sole filter and judge for an advanced hypothesis. Later on, philosophy (in its *logical positivist* version) took upon itself the furthering of this expulsion of anything non-empirically evidential from scientific discourse. This development rendered the very idea of metaphysics inimical to science—a prejudice to get rid of. Ultimately, behaviorism (a dominant trend in psychology at Wiener's time, whose reductionist dicta he was very aware of)[41] bracketed the mind as a metaphysical entity itself, focusing

---

[40] Rosenblueth et al. 1943.

[41] See Galison 1994, p. 243 (n). Wiener's disappointment with the poverty of "behaviorism" as a viewpoint in psychology is indicated above. Although the word "behavior"—and its derivatives—will be utilized often, it will no longer refer to the traditional psychological position. Instead, it will make a much broader referencing to the mode of being or process of coping of an entity—organic or artificial. More on this in the chapter concerning Ross Ashby's work.

instead on the inputs and outputs to a "black box". In other words, the very mention of teleology was grave heresy for the scientific and philosophical orthodoxy of the time—even if, in this case, what was labelled a "final cause" made no use of any metaphysical "force", merely referring to the goal to which the whole system aims, making use of corrective negative feedback. Still, Wiener went ahead and pushed for the use of the term, stating that "teleological" can perfectly be predicated of the behavior of a self-guided missile. As we will see later, while some saw in this revival attempt the sign of a science of the future, some others did not take it positively.

Once McCulloch went back to his home institution (the University of Illinois), he re-engaged into his experiments with renewed impetus, equipped with the providential insights from Wiener as faithfully conveyed by Rosenblueth. Norbert Wiener had met the psychiatrist McCulloch several years before, introduced by Rosenblueth at one of his Harvard evening meetings. McCulloch was already occupied with theorizing artificial neurons as early as 1927. The main idea, which had been haunting him for years, pivoted around the possibility of understanding the nervous system as a physical instantiation of a certain realm hitherto acknowledged as *a priori*, clear and untarnished: Logic. This pure realm stood in sharp contradistinction with the chaotic messiness of the human mind. However, both realms are somehow uneasily embedded in the brain. Thus, the enigma arose: How is it possible that a fallible and unpredictable entity, such as the human mind, could possibly grasp the exactitudes of mathematics? Where is the link that connects and fixes the "messy" with the "exact"? These questions were already present in McCulloch's early years as a college student, right in the midst of deciding whether to enter the theological seminary for becoming an ordained minister—faithful to his family's wishes.[42] His spiritual mentor at the time asked him about what he would like to do with his life. McCulloch told him that he would like to

---

[42] Puritan Protestantism ran deep in Warren McCulloch's ancestry. His father remarried a substantially younger southern lady who was deeply religious. Eventually, Warren's sister carried the religious family torch, becoming a committed Bible preacher. In fact, his ways strayed away from Puritanical Christianity in a manner that would ultimately carry consequences for his scientific endeavors, as we will see below when addressing the rift with Wiener (Heims 1991, pp. 31–32).

know "What is a number, that a man may know it, and man, that he may know a number?"[43] He recounts his intellectual pilgrimage as follows...

> I came, from a major interest in philosophy and mathematics, into psychology with the problem of how a thing like mathematics could ever arise-what sort of a thing it was. For that reason, I gradually shifted into psychology and thence, for the reason that I again and again failed to find the significant variables, I was forced into neurophysiology. The attempt to construct a theory in a field like this, so that it can be put to any verification, is tough.[44]

But the mathematical fleshing out of his developing insights was persistently evading him.

This wait came to an end, however, after McCulloch met someone whose genius might have been one of the most underreported biographical stories in the twentieth-century history of science.

Jerome Lettvin, a pre-med student who caught notice of the powerful medical experimental facility recently assigned to McCulloch, introduced the latter to the high school dropout Walter Pitts (1923–1969). Lettvin recounted that Pitts was once escaping from a group of bullies and had to hide in a library. While in his "refuge", Pitts stumbled upon a particular book, and once he began to read it, he felt drawn toward it. It was Bertrand Russell and Alfred North Whitehead's three volumes of the *Principia Mathematica*.[45] Pitts spent the subsequent week devouring the

---

[43] Warren McCulloch recalls the emergence of this question, which allegedly haunted him for the rest of his life, in the following manner:

> In the fall of 1917, I entered Haverford College with two strings to my bow—facility in Latin and a sure foundation in mathematics. I "honored" in the latter and was seduced by it. That winter [the Quaker philosopher] Rufus Jones called me in. "Warren," said he, "what is thee going to be?" And I said, "I don't know." "And what is thee going to do?" And again I said, "I have no idea; but there is one question I would like to answer: What is a number, that a man may know it, and a man, that he may know a number?" He smiled and said, "Friend, thee will be busy as long as thee lives." (McCulloch 1960, p. 1).

[44] Von Neumann 1951, p. 32.

[45] Whitehead and Russell 1927.

first tome, and found what he thought was an error in the work, triggering a letter to Russell. The British philosopher, impressed upon reading the missive, invited him to come to the UK to do graduate work by his side. Pitts was at the time 13 years old, and had no way of accepting such an offer. Two years later, Pitts—who was routinely beaten up by his own father, a plumber—finally escaped from his home in Detroit and headed to Chicago, in the hope of finding some intellectual comprehension and acceptance. At the time, Russell was visiting professor at the University of Chicago, and upon reconnecting with Pitts, he suggested the latter to attend the lectures of the logician Rudolph Carnap—which he did. Again, Pitts rather quickly pointed out to the logician what he thought were inconsistencies in a certain piece of his work—to the surprise of Carnap. He was 15 years old at this time, and the university took notice of the boy's powerful mind, routinely seen wandering around campus and sitting in on advanced graduate classes. Sensing the potential for an unusually bright future in academia, the university arranged for a modest amount of regular income for Pitts to literally survive. He was almost always homeless.[46]

When Lettvin introduced Pitts to McCulloch, the latter was profoundly impressed by his genius and soon practically adopted him, bringing both Pitts and Lettvin to live in his house. Not long after Walter Pitts had ironed out with passion the mathematical wrinkles that McCulloch's model was suffering from. The outcome of this team work was the 1943 seminal paper "A Logical Calculus of Ideas Immanent in Nervous Activity".[47] The proposal contained there pivoted around the notion of very simplified and idealized "neurons" heavily interconnected with each other, in such a way that after a massive conjoint operation—following the rules of Boolean logic—some behavioral outcomes in an organism could be accounted for. In other words, the mechanism taking place in animal cognition was subsumed under the realm of logical operations. The mind was underpinned by a real logical structure after all. It would seem that McCulloch was finally answering the haunting question of his youth. However, the paper failed to garner attention from its main target: psychologists and neurologists. Instead, an unlikely group of people took a

---

[46] Conway and Siegelman 2005, p. 115.
[47] McCulloch and Pitts 1943.

keen interest in it: The incipiently formed community of computer scientists. As we will see, this "productive misappropriation" would entail interesting consequences for the scientific enterprise that was about to be formed.

The mathematician John von Neumann (1903–1957), a Jewish Hungarian émigré who escaped to America fleeing the rising Nazi terror, was well acquainted with Wiener's work since at least a decade earlier. He was also a consultant for the Aberdeen Proving Ground where Wiener was working. Although Wiener's proposal for a computing machine delivered to Vannevar Bush was shelved at the time,[48] it propelled Wiener to be a consultant in computation for both the American Mathematical Society and the Mathematical Association of America.[49] Indeed the paths of both men were seemingly bound to intersect. In 1943, a key year for all these developments,[50] von Neumann joined the ultra-secret Manhattan Project, suddenly finding himself in need of a powerful computing machine that would crunch numbers with a complexity far beyond anti-aircraft firing tables. Mainly for this reason, he went to England to be acquainted with the work that the mathematician Alan Turing (1912–1954) was doing on computing machinery. When he came back, von Neumann confessed to being already obsessed with the enticing possibilities of computing power. Although he had access to Vannevar Bush's second analog computing machine, the 100-ton Rockefeller Differential Analyzer,[51] this machine was already overbooked for war-related tasks. Von Neumann thus became part of the consultant team for the improvement of Bush's successor computer, the ENIAC. Von Neumann was set to improve dramatically its shortcomings. McCulloch and Pitts' 1943 article was the model he used for his "First Draft" for the construction of still a more powerful successor, the EDVAC—whose design philosophy has prevailed until today under the name of "von Neumann architecture".[52]

---

[48] Mindell 2002, p. 275.

[49] Conway and Siegelman 2005, p. 145.

[50] Both foundational papers were published in 1943: "Behavior, purpose and teleology" and "A logical calculus of the ideas immanent in nervous activity".

[51] Bowles 1996, p. 8.

[52] The "von Neumann architecture", these days more commonly known as "stored-program computer", provided the advantage over the earlier computer

In this context, Wiener called for an exclusive meeting to be held at Princeton, where von Neumann was professor. The invitation was sent out in 1944 to seven people, including McCulloch, Pitts, Lorente de Nó, and Rosenblueth (who in turn could not attend this time). The meeting was co-organized by von Neumann and Howard Aiken, a Harvard mathematician. In 1945, the small "Teleological Society" meeting took place, and it served as an opportunity for von Neumann to reveal on the one hand the material feasibility of a working mechanical model of the brain, and on the other the state of computer science in general—at the time still a novel and fairly exotic area of knowledge. Wiener furthered his still forming views on the new science of communication. McCulloch and Lorente de Nó spoke about their model of the brain that was serving as a model for von Neumann's machine. The meeting was a success and it secured their desire to continue getting

machines in that the set of instructions could be stored as software in the machine—as opposed to physically changing hardware switches, as it was customary up until that time. The downside, however, is that since both instructions and data share the same bus (data communication device), performance could be compromised—effect known as the "von Neumann bottleneck". There were other differences that set it apart from the machine that Turing built in the UK. Martin Davis described it like this:

> Like the tape on that abstract device, the EDVAC would possess a storage capability—von Neumann called it "memory"—holding both data and coded instructions. In the interest of practicality, the EDVAC was to have an arithmetic component that could perform each basic operation of arithmetic...in a single step, whereas in Turing's original conception these operations would need to be built up in terms of primitive operations such as "move one square to the left." Whereas the ENIAC had performed its arithmetic operations on numbers represented in terms of the ten decimal digits, the EDVAC was to enjoy the simplicity made possible by binary notation. The EDVAC was also to contain a component exercising logical control by bringing instructions to be executed one at a time from the memory into the arithmetic component. This way to organize a computer has come to be known as the von Neumann architecture, and today's computers are for the most part still organized according to this very same basic plan . . . (Davis 2000, p. 182).

together on a regular basis. Von Neumann maintained close contact with Wiener from 1945 on. But the real launchpad for all these novel scientific approaches was just about to happen the following year.

THE MACY CONFERENCES

The war finally ended in 1945 and the time seemed appropriate to reconvene after the 1942 meeting. Frank Fremont-Smith was still in charge of the Macy Foundation's arm to fund medical sciences, and he promptly agreed, after McCulloch and Wiener's persuasion, to launch a recurring series of conferences. These series, for which they did not yet have a name, were to occur twice a year. The group was now richer, having merged both Macy's 1942 first list of attendants and the 1945 Princeton gathering of scientists and mathematicians. The original idea was to discuss theory at the meeting itself, and then test it in each of the attendants' own realms of practice, coming back later with a report. Subsequently, the general theory could be improved based upon the feedback from these reports. After the fourth conference the frequency of meetings was cut in half to just once a year—it was deemed more appropriate for giving sufficient time for implementation. The gatherings spanned the time period between 1946 and 1953 and they did not produce an official transcript until the sixth meeting in 1949. During the first conferences a mechanical device for audio recordings was used, but the results were far from satisfactory.[53]

The first conference was entitled "Feedback Mechanisms and Circular Causal Systems in Biological and Social Systems" and it took place in the Beekman Hotel, in New York.[54] The first lecture was delivered on the morning of March 8, 1946 by John von Neumann, joined by Lorente de Nó. They spoke about the purported correlations between the artificial networks of computing machinery and the mind's logical structures embedded in real neuronal wetware—realms just linked in 1943 courtesy of the McCulloch and Pitts' paradigm for the calculus

---

[53] "We tried some sessions with and some without recording, but nothing was printable." (McCulloch 1974, p. 12). See also Hayles 1999, p. 81.

[54] All of the conferences took place in the same hotel, except for the last one, which was moved to Princeton, New Jersey, in order to accommodate von Neumann— who nevertheless ended up not attending.

of neural activity. Von Neumann addressed the audience from the stand-point of the mathematician turned computer scientist, providing a panoramic landscape of the state of computer theory and construction —including his own. Lorente de Nó, as a neurophysiologist, correlated von Neumann's description of the logical processes in the metal with human physiology, pointing out possible instantiations where such out-look could fulfill an explanatory one—such as the yes/no character of neuronal firing. The proposal of an actual mechanism underlying mental processes, where a mechanical model of a realm that had forever escaped the grasp of science was being put forward, reportedly put the audience in a state of positive consternation. As with the 1942 conference, enthusias-tic discussion ensued—and this was the format kept throughout the rest of conferences to come.

In the afternoon, Norbert Wiener finally delivered himself what he could not do in person the year before at the science and mathematics gathering at Princeton. He gave a report on the new science that was being formed out of the realization of circular causality as a more complete tool for understanding goal-oriented systems—both natural and artificial. Wiener delivered a historical introduction of the evolution of mechanisms of control, from the time of Ancient Greece, going through Watson's mechanical governors and Maxwell's mathematical treatment of them, all the way to his "teleological machine" built (conjointly with Bigelow) for anti-aircraft weaponry. Rosenblueth was his lecture-partner. Faithful to Wiener's ideas as he was the previous year when delivering Wiener's keynote, Rosenblueth took care of pro-viding examples of the feedback mechanisms underlying the very survi-val of living systems. By this time, Wiener was more aware of the utmost importance of the notion of "communication" for the overall outlook of this emerging science—one that successfully encompassed both living organisms and artificial machinery:

[I]t had already become clear to Mr. Bigelow and myself that the pro-blems of control engineering and of communication engineering were inseparable, and that they centered not around the technique of electrical engineering but around the much more fundamental notion of the mes-sage, whether this should be transmitted by electrical, mechanical, or nervous means... [T]he group of scientists about Dr. Rosenblueth and myself had already become aware of the essential unity of the set of problems centering about communication, control, and statistical mechanics, whether in the machine or in living tissue.[55]

The second day, Warren McCulloch explained the proposal put forward in his cryptic article co-written with Pitts. As mentioned, this article failed to gain the attention of the audience to which it was originally intended—neurologists and psychologists—but in an ironic twist in the history of science, it served as the footprint upon which von Neumann designed a powerful computer machine—one whose fundamental structure remains pervasive to our day. The yes/no nature of neural firing, since it seemingly has a fairly direct correlation with the yes/no character of logical gateways, can provide a way to disentangle complex mental processes—as long as one is able to rigorously describe its smallest units in absolute detail. Von Neumann applied this thought to a physical contraption, giving it empirical instantiation, and his machine was born. McCulloch, on the other hand, was firmly on the side of finding an explanation for mental processes, and he thought that he found a paradigmatic one. The messiness of the human mind, however capable of dealing with the exactness of mathematics and logic could now be demystified: It is all a matter of complexity, which can ultimately be parsed into primigenial pieces of a network, each of them tractable in a yes/no fashion. Propositional calculus could provide a tool to understand nervous activity.

In all likelihood, half of the first two lectures of the first cybernetic conference were delivered with a Spanish accent—by Arturo Rosenblueth and Lorente de Nó.[56] The rest of the day was dedicated

---

[55] Wiener 1948a, pp. 8, 11. That was how Wiener recalled in his later book his earlier cybernetic experiences. The issue of communication would later become a contentious one, particularly regarding the issue of the possibility of transferring "meaning" over a mathematized articulation of communication—more amenable for exclusively syntactic conveyance. For more on this issue, see below in this chapter (8th Conference) and Chapter 5, Section "Machines can be Teleological"—including the notes.

[56] If one adds Santiago Ramón y Cajal to the duet, one could perceive a lineage of Spanish neuroscientific influence at the beginning of cybernetics (Ramón y Cajal's views on discrete neural structures inspired McCulloch's). This cybernetic "Spanish connection" is largely untreated in the literature. For Ramón y Cajal's influence on McCulloch, see Boden 2006, p. 188.

to the social sciences.[57] Once the conference was over, scientists went back to their own realms of research, invigorated by these novel ideas that promised to revolutionize science itself, largely by means of accurately expanding the realm of physics to areas heretofore considered inherently outside its grasp—such as life and mind (*pace* behaviorism, as we saw above).[58]

Wiener began to work with Rosenblueth on nervous activity in living flesh—commuting between Mexico City and Boston one semester every other year. He was trying to make sense of a problem he previously saw during his work involving gun turrets. As reported above, when the mounted gun received undamped positive feedback, it would engage in a wild swinging until some corrective (negative) feedback was applied. If his intuitions were correct, the same phenomenon of wild overcompensation observed in anti-aircraft machinery should take place in living organisms[59]— known in the latter case as *clonus*. Indeed, after Wiener begun experimenting with living animals, when an extra electrical impulse would be applied to a

---

[57] There are those who maintain that an important role of the Macy conferences was the spreading of cybernetic ideas to the wider realms of society and politics, particularly if one takes into consideration the insistence of Gregory Bateson at the first two Macy gatherings (1942 and 1946) for including the social sciences. Indeed, the social scientists present in the conferences were not unhappy about the great deal of media attention that the conferences began to garner. And there is also the famed case of Project Cybersyn: The democratically elected president of Chile Salvador Allende hired the cybernetician Stafford Beer in order to run a socialist Chile—before General Augusto Pinochet gave the *coup d'etat*. Be that as it may, the present writing is not concerned with the social extensions of the cybernetic proposal, but rather, with its core theoretical tenets, profoundly inter-twined with the nature of a machine. For an account inclusive of the social aspects of the conferences, see Heims 1991. For an account of the failed attempt to establish a cybernetic South American government, see Medina 2006, 2011.

[58] They might have even seen the "defense" of physics as one of the noble roles of the cybernetic mission. In fact, Dupuy wants to put more emphasis on the "extension of physics" aspect of the enterprise, somewhat against the notion of cybernetics as a "new science." However, not only are there authors who disagree, but also Wiener himself calls cybernetics a "new science" in non-vague terms. See Dupuy 2000, pp. 44–49; Conway and Siegelman 2005, p. 127; Wiener 1948a, pp. 14, 28.

[59] Wiener and Rosenblueth were already operating within a common framework for both animal and machine in no uncertain terms:

main nerve in a cat's leg, its limb would engage in spastic convulsions. Wiener and Rosenblueth were looking for a pattern. But results were evading them—and Wiener was vocal about his frustration. Working with living tissue greatly elevated the complexity of the study. After months of perseverance, however, they arrived to a startling finding. The increase of nervous activity responsible for the animal's limb movement was not additive, but logarithmic in nature[60]—not dissimilar with what he encountered when augmenting positive feedback during his experience with anti-aircraft systems. In other words, the clonic spastic movement observed in the cat's femoral muscle kept an exponential (not linear) correlation with the amount of electricity input in the animal's leg. The effect of the electric input seemed to be multiplicative, rather than additive, as it was the case in the swing and oscillating phenomenon found in gun turrets.[61] The significance of this

> We observed this pattern of contraction, paying attention to the physiological condition of the cat, the load on the muscle, the frequency of oscillation, the base level of the oscillation, and its amplitude. These we tried to analyze as we should analyze a mechanical or electrical system exhibiting the same pattern of hunting. We employed, for example, the methods of MacColl's book on servomechanisms (1948a, pp. 19–20).

[60] In the words of Wiener:

> ... the notions of facilitation and inhibition are much more nearly multiplicative than additive in nature. For example, a complete inhibition means a multiplication by zero, and a partial inhibition means a multiplication by a small quantity. It is these notions of inhibition and facilitation which have been used in the discussion ... Furthermore, the synapse is a coincidence-recorder, and the outgoing fiber is stimulated only if the number of incoming impulses in a small summation time exceeds a certain threshold. If this threshold is low enough in comparison with the full number of incoming synapses, the synaptic mechanism serves to multiply probabilities, and that it can be even an approximately linear link is possible only in a logarithmic system (Wiener 1948a, p. 20).

[61] Speaking of the significance of this logarithmic basis for overcompensating movements in both animal and machine, Wiener remarked:

> The most striking point is that on this logarithmic basis, and with data obtained from the conduction of single pulses through the various

finding was going to be spelled out shortly after. It meant reaching the zenith of Wiener's ideas regarding teleological mechanisms, since it tied in with research he had been involved in decades ago. There certainly was stuff to report in the next upcoming meeting.

The second conference, entitled "Teleological Mechanisms and Circular Causal Systems" took place in October 1946. Although it was mostly devoted to the possible social extensions of feedback mechanisms, it did have some science-related news. Walter Pitts, now an MIT student due to Wiener's influence, was to begin a Ph.D. tailored to Pitts alone—under the supervision of Wiener himself. The aim of the doctoral degree was going to be the construction of a three-dimensional model of neuronal structures— transcending the so far two-plane representation of neurons for a model more similar to the actual object, hoping that the functions of the neuron would be better captured and consequently understood. The degree of complexity sent waves of elation throughout the audience, especially among those well versed in mathematics, triggering McCulloch to declare that "the mathematics for the ordering of the [random] net has yet to be evolved".[62]

Right after this conference was over, another sub-conference started at the New York Academy of Sciences, and this one balanced things back to the "mathematics and science" side of things. Here Wiener presented to the attendees the conclusion to his evolution of ideas, tested under the fire of machine-building (the servomechanical anti-aircraft systems), and later, by his experiments on living tissue with Rosenblueth in Mexico.[63] Wiener had been noticing that there was a connection between what he was witnessing and thermo-dynamics. The "message" brought order to chaos—the default tendency in nature known as "entropy"—in both organic and inorganic realms. It would seem that beyond the general notion of the message, there lay the more fundamental notion of *information*. Information indeed "in-formed", giving shape to the realm in

> elements of the neuromuscular arc, we were able to obtain very fair approximations to the actual periods of clonic vibration, using the technique already developed by the servo engineers for the determination of the frequencies of hunting oscillations in feedback systems which have broken down (Wiener 1948a, p. 20).

[62] Conway and Siegelman 2005, p. 162.
[63] Wiener 1948b.

question, fighting off the primordial downward spiral of nature toward disorder and oblivion—the said entropy. Whenever there is information, entropy recedes—and vice versa. Entropy is the opposite of information, which in turn can be understood as negative entropy—or negentropy. Life is an informational force, negating entropy within the vastness of a merciless universe. Further, a working, "surviving" machine carries information that shapes its existence away from entropic decay.[64]

What Wiener was proposing was breathtaking, and the audience, by now reasonably adjusted to the excitement of these meetings, was once again in awe. McCulloch was sure to send copies of this talk's notes to several prominent scientists and mathematicians who were not present—among them, to the American mathematician Claude Shannon (1916–2001). Several years earlier, Shannon, a young MIT graduate and recently hired star employee at Bell Labs, was closely acquainted with Wiener and the war-related work of the latter—in a master-disciple kind of way. Picking up the theme shared here, but also as a culmination of long rationalizations regarding communication and information, he was to publish two years later (in 1948) the paper "A Mathematical Theory of Communication",[65] which stands to this day as a seminal document for "communication theory"—a field cross-listed between information and mathematics. Shannon became a regular attendee from the 7th conference on. The rest of the conferences will be alluded to throughout this book, some more than others, according to their relevance to this work. However, an outline of them follows.[66]

3rd Conference: "Teleological Mechanisms and Circular Causal Systems" (March 1947)

This conference was devoted mainly to psychology and psychiatry, somewhat to the displeasure of the members coming from the hard sciences. Von Neumann, having experienced a change of heart regarding

---

[64] For some possible theoretical extensions of Wiener's view on entropy, see Hayles 1999, pp. 100–108.

[65] Shannon 1948.

[66] A reconstruction of the first five conferences, without written record, have been put together mainly via personal recollections of some of the attendees. There have been two main attempts to historically reconstruct the cycle of conferences. First, historically, by Steve Heims (Heims 1991) and then thematically, by Jean-Pierre Dupuy (Dupuy 2000). There is also a summary that McCulloch wrote about the first three conferences in a letter written to Heinz von Foerster (McCulloch 1947).

the feasibility of using McCulloch and Pitts' networks to map the mind,[67] began to lobby for having the German biophysicist Max Delbrück invited to the meetings as a core member.

4th Conference: "Circular Causal and Feedback Mechanisms in Biological and Social Systems" (October 1947)

A further clash occurred between those members drawn to psychology and psychoanalysis, and those who were behind the more mathematical approach of McCulloch and Pitts' networks—in regards to mental activity. In particular, long-winded discussions ensued after the proposed distinction of Gestalt "forms", considered to be analogical, and McCulloch and Pitts' networks' inadequacy to capture them, due to the claimed digitality of the latter.[68]

5th Conference: "Circular Causal and Feedback Mechanisms in Biological and Social Systems" (Spring 1948)

Max Delbrück attended the meeting, to the joy of von Neumann—who, by this time, was seriously interested in genetic self-replication, and expected Delbrück to become part of the group core. Unlike any other meeting until the end, the entire first day was devoted to language. This bad timing might have played a role in Delbrück's disappointment. The second day Wiener spoke about chaos and Pitts about a formal modelling of chicken pecking. Delbrück never came back.[69]

[67] I will address this reversal in the chapter concerning von Neumann's contribution to cybernetics.

[68] In fact, the question regarding the possibility of finding the Gestalt "forms" of perception (shapes, sizes, patterns, etc.) was posed right at the first conference in 1946 by the neurologist-turned-psychoanalyst Heinrich Klüver. Some of the attendees were amicable to the plea for help, like Warren McCulloch—perhaps in his role of Chair of all ten of the Macy conferences. But some attendees found the whole issue unrigorous and ultimately intractable as a scientific issue—such as Walter Pitts, who was openly hostile to it right from the start. By this 4th conference, McCulloch also defined this position as squarely foreign to scientific discourse at best—and ultimately senseless at worst. By the 6th conference, McCulloch emerged very aggressively against the whole of psychoanalysis. By the 8th conference, Rosenblueth, having in mind McCulloch and Pitts' networks, asserted that mental events either occur, or they do not—a feature that leaves the unconscious memories of psychoanalysis, according to him, outside the realm of facticity and coherence.

[69] Heims 1991, p. 95.

The Hixon Symposium: "Cerebral Mechanisms in Behavior"[70] (September 1948)

A separate conference outside the Macy cycles took place that same year in Pasadena, at the California Institute of Technology (Caltech). This event was remarkable due to the fact that von Neumann now clearly attacked McCulloch and Pitts' networks—both McCulloch and Lorente de Nó were present. Von Neumann provided an alternative course of action for the cybernetic grand-enterprise: To shift focus toward the investigation of the possibility of constructing artificial cellular automata.[71]

6th Conference: "Cybernetics: Circular Causal and Feedback Mechanisms in Biological and Social Systems"[72] (March 1949)

Von Neumann, who could not attend the conference, sent a message to the attendees, which was read out loud. The audience took this missive seriously, and a further discussion regarding the inadequacy of McCulloch and Pitts' networks was triggered. This time, the issue was the insufficient capability of accounting for human behavior if relying on arrays of neurons alone—there were not enough of them. McCulloch defended its own model and Pitts eventually spotted an error in von Neumann's calculation regarding the quantity of neurons in a human brain, bringing the discussion to a halt.

The Austrian physicist Heinz von Foerster (1911–2002) was named editor-in-chief of the Macy conferences—partly in order to help improving his rudimentary English. He almost immediately proposed settling for one name for the conferences, *Cybernetics*, after Wiener's book, which was published the previous year. The acceptance of the proposal was unanimous. From that year on that title remained, and the previous mouthful was relegated to a subtitle.

7th Conference: "Cybernetics: Circular Causal and Feedback Mechanisms in Biological and Social Systems"[73] (March 1950)

---

[70] Published in 1951 under the same tittle (Jeffress 1951).

[71] This is the reason why the contemporary field of Artificial Life regards it as its birthplace. See Langton 1989.

[72] Von Foerster et al. 1950.

[73] Von Foerster et al. 1951.

Once again, lively discussion emerged regarding whether or not an analog model of the mind would be more suitable than a digital one—for example, the McCulloch and Pitts' model. No side seemed to get the upper hand. Also, a renewed attack against psychoanalysis as a coherent paradigm ensued, this time not only coming from the hard sciences (e.g., Pitts) but from the social sciences as well (e.g., Bateson).

Concern was voiced regarding unwarranted and overoptimistic accounts given by the popular media (e.g., *Scientific American, Popular Science*, etc.), when referring to the conferences as a "cohesive group" securely marching toward the mechanization of the human being. The environment of discussion, although exciting and lively, was far from embodying a "group mentality". McCulloch recalled an occasion in which Wiener was asking, while delivering a presentation, "May I finish my sentence?", to which an angry interlocutor replied "Don't stop me when I'm interrupting!"[74] In the same vein, a word of advice was thrown to those present coming from the social sciences: not to fall for the allure of the exactness of mathematics and the hard sciences, which seemed to have begun enjoying uncritical acceptance in their fields.

8th Conference: "Cybernetics: Circular Causal and Feedback Mechanisms in Biological and Social Systems"[75] (March 1951)

Both Wiener and von Neumann had by this time withdrawn from the conferences—for different reasons.[76] McCulloch remained the Chair, Pitts and Rosenblueth persevered and new people joined (e.g., Shannon). But the pair's absence was noticeable—particularly Wiener's.

---

[74] McCulloch gave further detail of the kind of milieu where the discussions were taking place:

> The first five meetings were intolerable...were not printable...Nothing that I have ever lived through...has ever been like...those meetings...You never have heard adult human beings, of such academic stature, use such language to attack each other. I have seen member after member depart in tears, and one never returned (McCulloch 1974, p. 12).

[75] Von Foerster et al. 1952.

[76] These reasons will be addressed in Chapter 3, Sections "The Decline" and "Traditional Explanations for the Collapse of Cybernetics".

The British physicist and Christian apologist Donald Mackay (1922–1987) criticized Shannon's views—well known by now. Shannon's position, where the message had to be stripped from any aspect of "meaning" in order to account for an accurate and engineeringly feasible conveyance,[77] would have, for Mackay, disastrous consequences for the proper understanding of the intended message.[78] Donald Mackay also entertained the possibility of constructing a learning machine with randomness factors embedded within, in order to account for a more human behavioral outcome. The American statistician Leonard Savage vigorously objected to it, claiming that adding randomness would help in no way to mimic human behavioral processes. Savage gave a presentation of his own, where he proposed a calculation procedure for decision-making based upon statistical factors. This time McCulloch objected to Savage, advancing the idea that the multilevel realm of decision-making is inherently contextual, and thus escapes the "linearity" of statistical analysis.

The American experimental psychologist Herbert Birch was aware of both Wiener's treatment of communication as homogeneously pervading the realms of animal and machine, and of Shannon's removal of semantics as essential for the rigorous treatment of communication. In face of this, he attempted to introduce a quality distinction.

[77] Boden 2006, p. 204. More on this in Chapter 5, Section "Machines can be Teleological", where the philosophical underpinnings of cybernetics are addressed.

[78] To be sure, Donald Mackay was not the only one concerned about this exorcization of semantics from a functional theory of communication:

> Soviet critics charged that Shannon's theory of communication reduced the human being to a "talking machine" and equated human speech with "just a 'flow' of purely conditional, symbolic 'information,' which does not differ in principle from digital data fed into a calculating machine." Wiener's formula, "information is information, not matter or energy," provoked a philosophical critique of the concept of information as a non-material entity. Repeating Lenin's criticism of some philosophical interpretations of relativity physics in the early twentieth century, Soviet authors castigated cyberneticians for replacing material processes with "pure" mathematical formulae and equations, in which "matter itself disappears." Cybernetics was labeled a "pseudo-science produced by science reactionaries and philosophizing ignoramuses, prisoners of idealism and metaphysics." (Mindell 2003, p. 82).

Coming from the area of research in animal communication itself, Birch differentiated the communication found in animals from the "real communication" found in humans. According to Birch, we speak of animal communication using a relaxed version of the notion of communication. The cyberneticists[79] reacted very negatively against this observation, alluding to the mentalisms and unwarranted metaphysical tenets underpinning such distinction. This occurrence is remarkable in its irony, given the event that was about to occur with Ross Ashby' presentation at the next conference.[80]

9th Conference: "Cybernetics: Circular Causal and Feedback Mechanisms in Biological and Social Systems"[81] (March 1952)

This was the only conference attended by the British medical doctor and psychiatrist William Ross Ashby (1903–1972) from the Ratio Club.[82] Ashby presented two papers where he proposed the change of behavior via random trials as an appropriate articulation of the phenomenon of learning in natural or artificial systems. The cyberneticists objected to this, in a fashion not in tune with the cybernetic mandate—to some extent borrowing Herbert Birch's context for argumentation alluded to above.[83]

Warren McCulloch voiced complaints about the governmental constraints gradually imposed upon several of the conference members. Indeed, the Cold War had already started and an environment of increasing secrecy let itself be felt. These remarks might have also served to deflect attention from the previous withdrawal of the core members (Wiener and von Neumann).

10th Conference: "Cybernetics: Circular Causal and Feedback Mechanisms in Biological and Social Systems"[84] (April 1953)

[79] Both "cyberneticist" and "cybernetician" are still used to refer to those who were part of the cybernetic project. Henceforth both terms will be used interchangeably.

[80] More on this in Chapter 6, where the contribution of William Ross Ashby is addressed.

[81] Von Foerster et al. 1953.

[82] More about the Ratio Club in Chapter 3, Section "The Ratio Club".

[83] More attention will be given to this theoretical reversal toward the end of the Chapter dealing with Ashby (6).

[84] Von Foerster et al. 1955.

THE MACY CONFERENCES 45

McCulloch acknowledged the decade of criticisms against his (and Pitts') model—particularly coming from von Neumann, who nevertheless built upon it arguably the most advanced computer of its time. McCulloch conceded that it is honorable to stay within the fine tradition of scientific refutability.

The discussion, despite its good humor, lacked "content" according to the German psychologist Hans Teuber, the assistant editor. Since he was just starting his academic career, in his view he had much to lose if such registration of events were to become public. Teuber was to offer his resignation if the transactions were to be published.[85] He later agreed to publish only the documents, without the ensuing discussions.

This was the only meeting that did not take place in New York. They moved it to Princeton, New Jersey, in order to accommodate von Neumann—who ended up not attending. But this is just half of the cybernetics' story . . .

[85] Hayles 1999, p. 75.

CHAPTER 3

Cybernetics: The Book, the Club, and the Decline

NORBERT WIENER'S *CYBERNETICS*

France might have been destined to be intertwined with the development of cybernetics right from its inception, as is shown by the fact that one of the first science historians who took notice of the movement was Pierre de Latil.[1] In that country both engineers and the intelligentsia took early notice of Wiener's ideas—while in America interest was more circumscribed to circles of engineering and the military, partly due to the ambience of secrecy owing to the Cold War.

In 1946, Wiener was invited to give a talk in Paris, at the *Collège de France*. After the talk, he went for a coffee with the Mexican-French publisher Enrique Freymann, and they hit it off—Wiener later spoke of him as one of the most interesting individuals he had ever met. Freymann asked Wiener if he would like to publish his ideas in a book. Wiener, surprised, answered that he could not think of anyone who would be interested in publishing them. After it was clear that Freymann was referring to himself, Wiener promised to submit a manuscript in three months. He subsequently went back to Boston, and then flew to Mexico City, to spend the period corresponding to his funded work with Rosenblueth. It is in this city that Wiener, now fluent in Spanish with a Latin accent, wrote

---

[1] See De Latil 1956.

© The Author(s) 2017
A. Malapi-Nelson, *The Nature of the Machine and the Collapse of Cybernetics*, Palgrave Studies in the Future of Humanity and its Successors, DOI 10.1007/978-3-319-54517-2_3

the entirety of the book—in the early hours of the morning, before heading to his laboratory work for the rest of the day with his Mexican colleague.

Wiener was on a roll; he wrote relentlessly. It seemed that he had been accumulating all these new ideas for the last years, and they were now boiling and eager to come out. Freymann received the write up in the mail after ninety days—much to his surprise, since he actually did not take seriously Wiener's agreement to deliver a manuscript promptly. He rushed to print out a first draft to send it back to Wiener for revision. In the meantime, MIT caught wind of Wiener's publication project and tried to persuade the publisher to concede the copyrights. To no avail. After intense negotiation, both publishers agreed to publish the book conjointly in the USA, but Freymann retained the rights for the French version (also in English) which was the one distributed internationally.[2]

In the book, after acknowledging with gratitude the influence and help received from Bush, McCulloch, Turing, von Neumann, Shannon, and others, Wiener did his due diligence in order to find a name for the new science he was advancing. Given that he was very well acquainted with the Greek and Latin classics—a legacy from his strict father, a Russian Jewish Harvard professor of philology—it did not take long for Wiener to engage in a thorough survey of possible terms for this novel discipline. At first, since "communication" lay at the core of the theory, he thought that the word *messenger* should be appropriate: *angelos* in Greek. However, the semantic overtones that this noun had historically acquired, that of an "angel" (*messenger* of the "good news"—*evangelion*) would obviously carry a connotation that Wiener was not trying to convey. So, he went on to explore the possibility of borrowing an image that had a long history of rich connotations for Greek thought, in fact used by Plato on more than one occasion: that of a "steersman"—*kubernetes*. The Ancient Greek κυβερνήτης, steersman or helmsman, is itself a construction formed out of two words: κυβερ (pilot) and ναυτος (ship). Plato used the term κυβερνητική in the *Georgias*, the *Laws* and *the Republic*, in order to refer to the "art of navigation" (or proper steering) of a community. Plato was referring to the political art of governance. Unbeknownst to Wiener at the time, the French physicist André-Marie Ampère did use "la cybernétique"

---

[2] Conway and Siegelman 2005, p. 175.

in 1834, referring also to the art of governance, but expanded to include elements of religion.[3]

Later on, the Latin derivation of the Greek word, *gubernator*, was acknowledged as the root of the English word *governor*, a device constructed by James Watt in the 1700s to be attached to steam engines for speed regulation—and later mathematically articulated. The British "natural philosopher" James Clerk Maxwell wrote in 1868 a mathematical treatise on these apparatuses—"On Governors".[4] This mechanical device is considered to have played an important role at the very core of the Industrial Revolution. Even if Maxwell did not use the term "cybernetics" per se, his 1867 treatise showed some prescient elements of control theory and was thus heralded by Wiener as the founding document for cybernetic thinking.[5] The mathematization of physical phenomena showing

---

[3] This is how Ampère defined "la cybernétique":

> Sans cesse il a à choisir entre diverses mesures celle qui est la plus propre à atteindre le but; et ce n'est que par l'étude approfondie et comparée des divers élémens que lui fournit, pour ce choix, la connaissance de tout ce qui est relatif à la nation qu'il régit, à son caractère, ses mœurs, ses opinions, son histoire, sa religion, ses moyens d'existence et de prospérité, son organisation et ses lois, qu'il peut se faire des règles générales de conduit, qui le guident dans chaque cas particulier. Ce n'est donc qu'après toutes les sciences qui s'occupent de ces divers objets qu'on doit placer celle dont il est ici question et que je nomme *Cybernétique*, du mot κυβερνητική, qui, pris d'abord, dans une acception restreinte, pour l'art de gouverner un vaisseau, reçut de l'usage, chez les Grecs même, la signification, tout autrement étendue, de l'*art de gouverner en général* (Ampère 1838, p. 141. Italics original).

[4] See Maxwell 1868.

[5] Wiener had this to say about the importance of feedback for cybernetic ancestry:

> We have decided to call the entire field of control and communication theory, whether in the machine or in the animal, by the name Cybernetics, which we form from the Greek κυβερνήτης or steersman. In choosing this term, we wish to recognize that the first significant paper on feedback mechanisms is an article on governors, which was published by Clerk Maxwell in 1868, and that governor is derived from a Latin corruption of κυβερνήτης. We also wish to refer to the fact that the

corrective negative feedback indeed lay at the core of cybernetics. Gottfried Wilhelm Leibnitz (1646–1716), due to his idea of a universal language of calculus, was proclaimed by Wiener as cybernetics' patron saint.[6] Such illustrious lineage was reason enough for Wiener to coin *cybernetics* as the word that would name from now on his newly unveiled science.

It would seem fair to say that the uniqueness of Wiener's book relies on the fact that he managed to put under one cover many connected ideas regarding communication and information that were current on both sides of the Atlantic—mainly due to war-related research, where the USA officially invested. In fact, the British hub of cyberneticists, the *Ratio Club*,[7] entertained the implication that such was the reason for Wiener's in particular, and America's in general, lead in the new field. Wiener had come up with a perfectly timed book that not only gave these ideas wide circulation, but also served as a hub for the cybernetic insights already floating around. Indeed, these ideas were emerging in several parts of the world, including France, Sweden, the Soviet Union, and the UK.

steering engines of a ship are indeed one of the earliest and best-developed forms of feedback mechanisms (Wiener 1948a, pp. 11–12).

[6] Wiener wrote the following about Leibnitz as the "spiritual father" of cybernetics:

If I were to choose a patron saint for cybernetics out of the history of science, I should have to choose Leibniz. The philosophy of Leibniz centers about two closely related concepts—that of a universal symbolism and that of a calculus of reasoning. From these are descended the mathematical notation and the symbolic logic of the present day. Now, just as the calculus of arithmetic lends itself to a mechanization progressing through the abacus and the desk computing machine to the ultra-rapid computing machines of the present day, so the *calculus ratiocinator* of Leibniz contains the germs of the *machina ratiocinatrix*, the reasoning machine. Indeed, Leibniz himself, like his predecessor Pascal, was interested in the construction of computing machines in the metal. It is, therefore, not in the least surprising that the same intellectual impulse which has led to the development of mathematical logic has at the same time led to the ideal or actual mechanization of processes of thought. (Wiener 1948a, p. 12. Italics original.).

[7] More on this group of scientists below.

War-related research was the main motivation for all of them. The British and the rest, without official support, missed that opportunity. Wiener, who had formal government back up, seized it.

One could safely assume, however, that Wiener himself had the main cybernetic ideas boiling and looking for a release valve. For starters, he already had three written pieces amounting to a solid cybernetic core: The top-secret "Yellow Peril" of 1942,[8] his co-authored 1943 article on teleological mechanisms,[9] and the 1946 lecture at the New York Academy of Sciences.[10] Wiener also started with the acknowledgment of the need for a common vocabulary for the emerging discipline, after witnessing the slippery slope of borrowing terms from one field to another, hence imposing one worldview onto the other and missing the substance of the novel approach.

Wiener's newly found insights, regarding the poverty of the notion of behaviorism after his experience with the predictor, were spelled out. The utmost importance of negative feedback, after Rosenblueth's input from physiology, was carefully articulated. The existence of a common process aimed at survival, common to both animal (homeostasis) and machine (self-regulation) was established: Both were teleological mechanisms. Wiener's startling realization of the defining importance of "information", as the carrier of order, into and against the natural tendency toward chaos in reality—entropy—was given due treatment. Beyond control theory and communication, information (or "negative entropy") was the way nature regulated itself, in the organic and inorganic, as already hinted in the discipline of physics by the second law of thermodynamics—an area with which he was intimately acquainted, from decades ago, due to his studies of chaotic behavior. Reality's inherent tendency toward decay is counterbalanced by information, "life" being a paradigmatic example—but also a machine, as an island of will surviving in the midst of, and against, chaos.

Wiener provided cases which, linked via mathematical treatment, convincingly showed that the notions referred to above equally applied to both the living and the non-living—and that a common methodology of

[8] See Wiener 1949.
[9] See Rosenblueth et al., 1943.
[10] See Wiener 1948b.

study and experimentation was now emerging.[11] For example, he linked the case of a machine breaking down due to an unrestrained overflow of positive feedback with that of a brain overwhelmed with information when no more network storage is available. In both cases the lack of a regulative negative feedback would result in disastrous consequences—in the former one due to a lack of balance that breaks the system (probably a perennial loop), in the latter due to a circular (vicious) reasoning that results in neurosis, and eventually, insanity.

Wiener also ruminated about the social consequences of the new science, and the subsequent moral responsibility for the cybernetician. Up to that moment he was keeping those thoughts to himself. Two events, one past and one ongoing, may have triggered his cautious view. Firstly, the dropping of two nuclear bombs that killed more than 150,000 people—mostly civilians. Secondly, the growing ambience of secrecy that was effectively crippling the otherwise free flow of communication between scientists—part and parcel of the very essence of scientific practice, in Wiener's view. Expanding on the foreseeable consequences of a dramatically augmented capacity for pervasive control, he predicted a social and economic change that could be regarded as disturbingly prescient. Wiener saw an ugly head gradually rearing out of the newly found capacity of communication engineering that does not distinguish between man and machine.

"Information is information, not matter or energy",[12] Wiener would say. Even less it would be the "content" of a message. The delivery of order would be, quite literally, the most powerful force in the universe: The very impetus that fights against reality's own downward spiral into oblivion—the latter having been reason for hopelessness in more than one scientific mind. In a "cybernetically organized mankind"[13] the manipulation of information, now understood and appropriated, would have the pervasive capacity of profoundly and substantially modifying the physical and the non-physical, bringing the capacity of reality-control and modification to a level never seen in history. If the *industrial* revolution downgraded the "human arm by the

---

[11] Johnston 2008, p. 29.

[12] Wiener 1948a, p. 132.

[13] This is Martin Heidegger's term. The German philosopher was closely following the developments of cybernetics. For him, it embodied the culmination of the unfolding of the history of metaphysics—or the essence of technology. More on this in the next chapter.

competition of machinery",[14] the *cybernetic* revolution "is similarly bound to devalue the human brain, at least in its simpler and more routine decisions".[15] And that would be just the beginning.

The scientist or engineer vaguely familiar with cybernetic ideas was certainly expected to have a "field trip" of enjoyment in reading the piece. For one, engineers finally made the connection, only perceived before by vague intuition, between what they were doing at the level of electronic signals transmitted via networks (e.g., telephone), and the nature of servomechanisms at large. There was no articulated theory of feedback that would equally apply to noise channels (in order to guarantee the accurate conveyance of a message) and to industrial machinery (in need of self-regulating devices for smooth behavior). Let alone the startling demonstration that such regulating circular causality had always been present in nature itself (e.g., homeostasis). Several areas of science took notice at roughly the same time, while various engineering departments rushed to formalize the adoption of these views—which were not entirely new, but for the first time articulated with precision.[16]

Wiener's book was riddled, after the passionately written "Introduction", with complex equations—which in the first edition turned out to contain some mistakes.[17] However, it also contained powerful statements, sometimes slightly aphoristic, which revealed a mind that had been occupied for years in the developing of these notions. Some reviewers dismissed it, partly due to its complex mathematics and partly due to the sometimes non-linear way of argumentation. But what happened not long after its publication by far exceeded everyone's expectations, not only the author's and the publisher's, but also the "cybernetic group"[18] of Macy at large. *Cybernetics: Or Control and Communication in*

---

[14] Wiener 1948a, p. 27.

[15] Ibid.

[16] Conway and Siegelman 2005, p. 184.

[17] Wiener blamed Pitts for having misplaced the corrected manuscript for a while. This was a source of uncomfortable tension between them. Once the document reappeared, it turns out that someone mistakenly swapped the corrected version with the original one, sending the unchecked one back to France. This is the one that was printed and published—carrying some errors (Conway 205, pp. 203–204).

[18] This is Steve Heims' coinage. See the title of Heims 1991.

the Animal and the Machine became a best-seller on both sides of the Atlantic. The readership went well beyond those human groups circumscribed to mathematics, engineering or hard sciences. The social and political scientist, the philosopher and the humanities scholar, the middle-class intelligentsia, the entrepreneur, and businessman, the typical university educated layman at large...all began to read *Cybernetics.*

Finally, this legendary group led by this mysterious genius mathematician was producing a manifesto relatively accessible to a wider audience. The Introduction to the book is an enthusiastic narrative that can give a very good idea to any reasonably intelligent reader of what the project was all about. The popular media *ipso facto* noticed the shockwave and global buzz, and Wiener soon was the subject of articles in such magazines as *Scientific American, Newsweek,* and *Time.* Later this wave was joined by *Business Week, The New York Times,* and *The Times Book Review.* Later on, Wiener was portrayed in *Life* and *The New Yorker,* and *Time* had a cover story dedicated to him. Suddenly he was being invited by the most prestigious universities and societies for keynote addresses—for example, the 1950 International Congress of Mathematicians, at Harvard University.

Cybernetics was having a palpable impact in science, philosophy and beyond on a global scale, to the point of extending its influence to that part of the world that was, due to the Cold War, regarded as being outside Western reach: The Soviet Union.[19] Exchanges between cyberneticians coming from both the political, military, and economic rival sides of the globe were relatively fruitful, indeed to the extent of raising some eyebrows among surveillance authorities on both sides. Interestingly (albeit perhaps expectedly, due to socialism's canonical attraction toward all things "control"), the Soviet cyberneticians counted on the tacit but firm support of the state,[20] in sharp contrast with their British counterparts, who had to rely mostly on the practitioner's own economic means and free time.

---

[19] Norbert Wiener engaged in active professional exchange with his Soviet colleagues when it came to cybernetic research, scientific insights, and collaborative conferences. In the Soviet Union, cybernetics enjoyed a form of *placet* from the ruling Communist party, probably due to its aura of "control." These rapprochements were expectably not celebrated by governments of either side of the "Iron Curtain." (Conway and Siegelman 2005, ch. 16).

[20] Ibid. For the Soviet development of cybernetics—a story still fairly unknown in the West—see Gerovitch 2002.

Within the West, cybernetics' reach trascended the English-speaking world, overcoming the centuries-old multi-level historic animosity between Continental Europe and its Anglo-American occidental partner. Indeed, as an enterprise whose core practitioners exclusively used English as their language of communication and practice, cybernetics found enthusiastic reception in France (e.g., Lacan[21]) and in Germany (e.g., Heidegger[22]). As mentioned, the first international edition of *Cybernetics*, to the chagrin of Wiener's own home institution printing press (MIT Press), was published by a French publishing house—following up the conferences in France to which he was invited.

Furthermore, it was one of the few scientific movements in modern history that managed to garner widespread fame—indeed well beyond scientific and academic circles. Mainstream media was very receptive to cybernetic developments, arguably due to the bold proposals (for the time) of understanding mind as machine and life as mechanical. Reasons for this existence in the spotlight can also be found, however, in the aura formed around the visually engaging spectacle that cybernetic robots used to provide—a veritable "gallery of monsters".[23] This almost surreal world—a pictorial reminder of the post-modern times that were being lived—was filled with reported anecdotes of autonomous robots chasing women's legs,[24] and children being surprisingly (again, for the time) prone to interact with these "mechanical animals".[25]

Here it is relevant to indicate that the role of Britain in cybernetics emerges as fairly substantial when we refer to these mechanical

---

[21] For an account of Lacan's appropriation of cybernetic thinking applied to psychoanalysis, see Johnston 2008, ch. 2.

[22] See Heidegger 1981 and 1977b.

[23] This is Andrew Pickering's coinage. Cf. Pickering 2005.

[24] This reportedly happened with the phototropic tortoises. The bright type of stockings used by women at the time used to reflect light. Spotted by the machine, it would resolutely aim toward the feminine lower limbs while avoiding men's trousered legs (Hayward 2001, p. 624).

[25] De Latil's book shows a photograph of Grey Walter, his wife and their son playing with a "tortoise". The description below reads: "Vivian Dovey and Grey Walter have two offspring: Timothy, a human baby and Elsie, the tortoise, of coils and electronic valves. Timothy is very friendly with his mechanized sister." (De Latil 1956, p. 34).

contraptions. The British collaboration with the cybernetic movement has been lately acknowledged with increased emphasis.[26] To be fair, despite Wiener's somewhat belittling remarks after a visit to the UK,[27] some members coming from across the Atlantic did play an important role in the unfolding of the whole cybernetic enterprise—in particular toward the end. In line with the more informal cybernetic development coming from British soil, they had their own scaled down version of the reportedly fancy Macy gatherings[28]—indeed, in the form of a small but exclusive "club".

THE RATIO CLUB

The Ratio Club was a small community of British scientists that gathered in London on a regular basis to discuss all things cybernetic. Just as the Macy meetings, it served as a hub where scientists gathered to share and discuss their insights, cognizant that there was no other forum at the time where they could discuss cybernetics. Expectably, they were filled with enthusiasm, and members would always be looking forward to the next meeting. Nevertheless, there were some remarkable differences from the Macy conferences as well. Although the Macy gatherings appreciated the spontaneous discussions that followed the presentations, the London meetings had a decided bent on keeping the gatherings more informal than in New York. The pace was set by talking over drinks, with no records ever produced of the discussions. They were to meet once a month, at

[26] First by Husbands and Holland (2008); later by Pickering (2010).

[27] Wiener characterized his 1947 visit to England thus:

> I found the interest in cybernetics about as great and well informed in England as in the USA, and the engineering work excellent, though of course limited by the smaller funds available ... I did not find, however, that as much progress had been made in unifying the subject and in pulling the various threads of research together as we had made at home in the States (Wiener 1948a, p. 23).

[28] Conway & Siegelman, in describing the first Macy meeting, wrote that "They gathered for two days and two nights around a great circle of tables at the Beekman Hotel on Manhattan's Upper East Side, with their lodging, meals, and the cocktails Wiener would not touch paid for by the Macy Foundation." (Conway and Siegelman 2005, p. 155).

night, in the basement of a hospital—arranged with chairs, food, and beer. These kept going in a relatively uninterrupted fashion from 1949 to 1955. In order to maintain the discussion in total freedom, without affectation from an "authority figure", no professors were allowed. Also, if someone would become someone else's boss, it was expected he would resign. In this spirit, the one-time proposal for a follow up in a more professional setting, probably with a journal, was flatly rejected.[29]

The origin of its name can be traced back to the semantic implications stemming from the fact that the Latin root *ratio* underpinned several concepts important for the group.[30] Leibnitz was regarded by Wiener as a precursor to cybernetic thinking, when the latter acknowledged that his *machina rationatrix* was anteceded by Leibnitz' *calculus rationator*.[31] The club members were also interested in the notions of rationarium (statistical account) and ratiocinatius (argumentation). *Ratio*, the common root for all these notions, seemed like the right choice.

Over and above the mentioned differences existing between the Ratio gatherings and Macy's, there was a feature that set them apart in a qualitatively distinct way. Despite the highly abstract origin of its name, the club was largely about device making, and the reason for this lay partly in its recent history—indeed a sad history that for some of the members was still very fresh. One member died just before making it to the first meeting: Kenneth Craik (1914–1945). Craik's "synthetic philosophy", which will be addressed in the next chapter, strongly informed and defined an essential aspect of the group: Its emphasis was on device-construction as a means of explanation. In fact, at least one Ratio member referred to what was going on in the American side as "thinking on very much the same lines as Kenneth Craik did, but with much less sparkle and humour".[32]

---

[29] This was Ross Ashby's idea. He was more than amenable to the prospective of constituting a more formal, open society that could reverberate into a journal. Other members vehemently disagreed. (Husbands and Holland 2008, p. 128).

[30] *Ratiocinatio* is St. Thomas Aquinas' term for discursive reasoning, namely, the natural reasoning used by humans when a truth is sought after—as opposed to the angels' or God's intuitive knowledge of things (Husbands and Holland 2008, p. 102).

[31] See above note on cybernetics' patron saint according to Wiener.

[32] Husbands and Holland 2008, p. 14.

One main criterion for recruiting members (which were always kept constant at around twenty, including core members and guests) was that they should have had, in John Bates words, "Wiener's ideas before Wiener's book appeared".[33] There was substantial emphasis placed on this mandate, since there was the conviction that the American side, to a great extent, was enjoying the fruits of having merely put cybernetic ideas together with impeccable timing. As advanced above, many of those ideas were allegedly already floating around before Wiener and company—and adding insult to injury, many were coming from Britain. In fact, the Ratio group did have prominent members that went down in history as prominent cyberneticians or thinkers in their own right,[34] such as Alan Turing, Grey Walter,[35] Donald Mackay, and Ross Ashby.

These almost patriotic statements might, up to some point, be understood as reactive to some of Wiener's remarks regarding the state of cybernetics in England. As mentioned above, Wiener acknowledged in his book the originality and enthusiasm that British scientists demonstrate, but pointed out, probably not without reason, that they were not as developed and organized as in America.[36] After all, the very nature of

---

[33] This criterion was several times repeated in letters throughout the formation of the club. The neurologist John Bates, who came up first with the idea of these gatherings, wrote in one of his invitation letters: "I know personally about 15 people who had Wiener's ideas before Wiener's book appeared and who are more or less concerned with them in their present work and who I think would come." (Husbands and Holland 2012, p. 242).

[34] They did acknowledge as "honorary members" McCulloch, Pitts, Shannon and either Wiener or Weaver.

A list of the club membership in a member's correspondence reads "Mc", "P", "S" and a letter that could be either a "U" or a "W". Husbands speculates that it could refer to either Norbert Wiener or Warren Weaver (Husbands and Holland 2008, p. 98).

[35] William Grey Walter (1910–1977) was an American (naturalized British) roboticist and neurophysiologist, largely self-taught while working at hospitals. He is best remembered by his famous mechanical "tortoises", displaying animal-like behavior at festivals and media gatherings. He called them both *machina speculatrix* (and each, Elmer and Elsie) because the way in which they would find their way around environments seemed to suggest that they were "deciding" about possible routes.

[36] Wiener 1948a, p. 23.

the Ratio Club pointed to a hobby-style, "indie" type of pursuing of cybernetic insights,[37] without an official backup coming from the British government.[38] Wiener's remarks might have been exacerbated in their effects after McCulloch's talk at the Ratio Club. There was a unanimous feeling that it fell flat. In addition, due to McCulloch's well-known penchant for flowery language and grandiose remarks—after all, he was heavily trained in the Classics—a member voiced the feeling that they found "Americans less clever than they appear to think themselves".[39] Alan Turing referred to him as "a charlatan" straight up.[40]

In fact, there seemed to be an undercurrent of wounded pride throughout the British cybernetic development.[41] As an example, J. O. Wisdom's article "The Hypothesis of Cybernetics" might fall within this context.[42] Wisdom was intrigued by both the audacity of the cybernetic proposals, and the rigor—even elegance—of their theoretical premises and procedures. However, Wisdom somewhat reduced the cybernetic hypothesis to a sophisticated mechanistic view of the nervous system. And the reason for this might be partly historical in nature.

Not long after the role of negative feedback in machinery had been carefully examined, the hypothesis that claimed that the nervous system was, in fact, a machine with negative feedback, saw the light. This might have been influenced by the fact that, particularly in the UK, many biology researchers were recruited, due to the war efforts, to work in radar research

---

[37] Pickering 2010, p. 10.

[38] In fact, the opposite could be said as having been the case regarding the support of cybernetic pursuits by the British government. As a telling example, at some point British Intelligence begun to be concerned about the amount of military surplus that Grey Walter was purchasing, and instructed someone to find out, in a covert but official capacity. Walter was just scavenging for useful parts to build his "tortoises" (Husbands and Holland 2008, p. 127).

[39] Husbands and Holland 2008, p. 114.

[40] Hodges 1983, p. 411.

[41] Perhaps present until today's accounts of what happened back then. British philosopher Andrew Pickering does not hide that he wants to "set the record straight" regarding the absence in history of the British contribution to cybernetics. In fact, he is clear in stating that his book is not a treatment of cybernetics, but of an aspect of British cybernetics (Pickering 2010, pp. 3–4).

[42] See Wisdom 1951.

and development.[43] These biologists became thus familiarized with electronics and mechanics, and biological realms began to be framed under a mechanistic guise. Conversely, urgently needed research on radar technology was an incentive to think of it as an artificial extension of human sensoria. This British emphasis on radar development, as seen in the previous chapter regarding American predictors being coupled with English radars, was an important aspect of the context of development of the Ratio Club.

Wisdom was ultimately critical of the scope and limits of the cybernetic enterprise, equating it to a mechanical hypothesis of the nervous system—which was but just *one* aspect, even if important, of the cybernetic epistemology. Wisdom seemed to have missed the crux of the theoretical pillars of the enterprise—namely, that of the assumption of the perfect possibility, nay, the hitherto unacknowledged existence, of an entire class of non-material (disembodied) machines. And the patriotic context mentioned above may have had something to do with this. It could be reasonably entertained that Wisdom might have wished to point out that the tradition of research on the nervous system within the framework of computation was on the British side at least as good as on the American one—as depicted by von Neumann and McCulloch.

Be that as it may, the group enjoyed a frequency that Macy, only once a year, did not. Also, at least reportedly, the meetings were less messy and conflictual than at Macy—as reported prior. Rather, members acknowledged that a true interdisciplinary discussion (at least in spirit) used to emerge.[44] Most remembered the Ratio meetings as bringing intense

---

[43] John Bates described the core membership of the club as "half primarily physiologists though with 'electrical leanings' and half communication theory and ex-radar folk with biological leanings" (Pickering 2010, p. 59).

[44] As opposed to the chaos of Macy where some—mostly social scientists—used to be "lost". This is how Jean Pierre Dupuy describes it:

> The cyberneticians showed no reservation about entering into technical discussion with the widest variety of specialists, examining with equal ease the results of a psychoacoustic experiment and a theory of the conditions under which the words of a language acquire a specific individual meaning. The few generalists present, the most notable of whom

academic and intellectual fulfillment to its members—in fact, for some, those meetings gave the most insightful and helpful insights for their own, later prominent careers.[45] From 1949 to 1951, the meetings were as frequent as scheduled; however, attendance begun to dwindle after 1952. There were only four meetings in 1953, three in 1954 and two in 1955. There was one more, sort of "final reunion" in 1958. And the club never met again.[46]

The Decline

From the pre-Macy conferences up until half the lifespan of the Macy cycle, most cyberneticians were in a state of a nearly constant adrenaline rush. But some began to grow worried regarding the grand-promises that this new science was to deliver to the world, as reported by the popular media. This conflicting reception can be noticed when seeing from an eagle's view point both the unanimous motion of approval for having the series of Macy conferences' title changed to Cybernetics *and* the words of cautious advice voiced by its own members.

Despite being the target of popular attention (or perhaps, precisely due to it), Norbert Wiener was right from the beginning relentless in

---

was Gregory Bateson, often found themselves lost. In their frustration at being unable to follow the discussion, they were apt to beg their collea- gues not to lose sight of the universalist vocation of cybernetics. Careful examination of the transactions of the conferences makes it painfully clear how "out of it" Bateson actually was (Dupuy 2000, p. 88).

[45] Husbands and Holland 2008, pp. 138–141; Holland and Husbands 2011, pp. 120–122.

[46] Reasons for its slow decline were diverse. For one thing, it was too far for many, having to return home very late at night, to the annoyance of their respective wives. On the other hand, it was fairly expensive to get there, and for some the costs were not covered by their institutions. However, it would seem that the most important factor was that by the early 1950s cybernetics had already reached a fairly mainstream status, and so the ambience of originality, uniqueness, and to some point compelling mystery and scientific heroism was no longer there. Allegedly members lost interest as a consequence (Husbands and Holland 2008, p. 129). Needless to say, these are just some of the reasons as reckoned by Husbands and Holland. There might be others that shall be gradually unveiled later in time, after researchers take interest in the fate of this interesting group.

trying to calm down what he regarded as an unwarranted hype around the cybernetic project. Not only would he shun claims that could be deemed as plain excessive, but indeed more importantly, he was particularly skeptical of the social applications of the powerful cybernetic concepts. Bridging heretofore ontologically incompatible realms, such as man, machine, and life—thus extending the realm of physical science—was difficult enough. And Wiener felt understandably proud about such a feat. But extending the mechanical umbrella to portions of reality where the amount of variables was qualitatively over and above the tractable was the place where he would draw the line. The unpredictability of social behavior, for Wiener, inherently escaped numerical framing.[47]

These are relevant observations, since they point to a web of internal tensions already present within the project right from the start—tensions that have been traditionally highlighted as part and parcel of the eventual fate of cybernetics. Almost right from the beginning, the Macy conferences witnessed sociologists and anthropologists as some of the most enthusiastic "converts" to the cybernetic view—to the discreet discomfort of some cyberneticians, Wiener being prominent among them.[48] Seemingly, some scholars emanating from the "soft"

---

[47] While Norbert Wiener was voicing his cautious outlook on applying a cybernetic framework to society, just a year after his seminal 1943 paper, fellow cybernetician John von Neumann was publishing a book with Oskar Morgenstern on *Theory of Games and Economic Behavior* (1944), which launched within economics—itself traditionally regarded as a social science—entire subfields heavily based upon mathematics. There are those who believe that von Neumann contributed to the theoretical backbone of contemporary economics more than what he is usually credited for. For an instance of this view, see Mirowski 2002.

[48] This is how Wiener would characterize the import of cybernetics over the social sciences:

> I mention this matter because of the considerable, and I think false, hopes which some of my friends have built for the social efficacy of whatever new ways of thinking this book may contain. They are certain that our control over our material environment has far outgrown our control over our social environment and our understanding thereof. Therefore, they consider that the main task of the immediate future is to extend to the fields of anthropology, of sociology, of economics, the

sciences were more than eager to adopt a heuristic tool that would bring the scientific validation that was eluding them right from the foundation of their own fields—including psychology, *pace* Skinner.[49]

The tension witnessed between these two grand realms of knowledge within cybernetics may well point to another source of unease: the radical interdisciplinarity of the movement. Part of the allure of cybernetics was the deep realization, common across fields, that this neo-mechanistic understanding of life and mind could encompass a novel unifying language, structured upon the strengths of physics and mathematics. This approach, which necessarily entailed the interdisciplinary gatherings of Macy, was for some members one of the most attractive features of the cybernetic proposal.

However, for some others, particularly coming from the "hard" sciences, the attempt at dialoguing with other disciplines was coming at a high price, entailing a substantial dissolving of their hard earned, specialized, field-centric semantics—and all that just for hooking up with not-so-neighboring fields anyway. This alleged theoretical bridge would be founded, according to its critics, more on vagueness than on substance—and that was too much of a hard pill to swallow. Some indeed never came back.[50] All tensions and early defections notwithstanding, the overlapping consensus among cyberneticians was that something deeply important—indeed, a type of 2nd Scientific Revolution of sorts—was in the making, and thus they chose to be patient and stick to the cybernetic project. The opportunity for a profoundly positive outcome for science would be too great to dismiss, just due to some wrinkles in the way it was being laid out.

methods of the natural sciences, in the hope of achieving a like measure of success in the social fields. From believing this necessary, they come to believe it possible. In this, I maintain, they show an excessive optimism, and a misunderstanding of the nature of all scientific achievement (Wiener 1948a, p. 162).

[49] See the first section of the previous chapter (Chapter 2. Section "The Problem of Anti-aircraft Weaponry") for Wiener's complain against the poverty of traditional behaviorism.

[50] One striking example of this attitude was Max Delbrück's refusal to return to the conferences, mentioned in the previous chapter (2). I will come back to this occurrence in the chapter pertaining to von Neumann's work (7).

Probably most importantly, and this aspect has arguably not enjoyed its due consideration, is the impact that cybernetics had on its members, underpinning later advances in technology and science. In the case of American cybernetics, the fields of information theory, computer science, genetics, artificial intelligence (AI) and artificial life, could be traced back to its cybernetic roots in a fairly straightforward manner—although this uncompleted task has not received enough attention from the scholarship.[51] In the case of the Ratio Club, the testimony of some of its members regarding their indebtedness to cybernetics is particularly striking.[52] The difficulty in pinning down what exactly it is about cybernetics that manages to carry until now an aura of fascination might be deeply intertwined with this holistic and fluid aspect of its scientific ethos. All these historical circumstances, theoretical features and even spin-offs emanating from the cybernetic enterprise, stand as witness to the greatness of the project

[51] In fact, Gualtiero Piccinini's review of Jean Pierre Dupuy (2000) points to this lack, fulfillment allegedly somewhat promised in the book's preface (Piccinini 2002). Still Dupuy (2000, ch. 2) could be a good start for connecting cybernetics with the cases of artificial intelligence and cognitive science. The legacy of cybernetics in the areas of genetics and information theory is briefly covered by Kay (2000, ch. 3). Langton (1989) addresses von Neumann's foundational role for artificial life. For the cybernetic ancestry of computer science, see Boden (2006, pp. 157–162, 195–198).

[52] It would seem that the strong emphasis on the informality of the club was precisely what made some of its members cherish it even after its extinction. Donald Arthur Sholl, a member of the club whose research in neurology is consulted until now (Sholl 1956) said this of the impact of the discussions in his professional life:

> I consider membership of the Club not only as one of my more pleasant activities but as one of the most important factors in the development of my work. I have stressed before how valuable I find the informality and spontaneity of our discussion and the fact that one does not have to be on one's guard when any issue is being argued. At the present time we have a group of workers, each with some specialized knowledge and I believe that the free interchange of ideas which has been so happily achieved and which, indeed, was the basis for the founding of the Club, largely results from the fact that questions of academic status do not arise (Husbands and Holland 2008, p. 129).

according to its founders, its profound role displayed upon its members, and its well-acknowledged influence at the time on the scientific community at large.

However, after an intense decade of enthusiastic interaction between scientists, mathematicians and engineers of the first order—reciprocally connecting in one novel theoretical structure of common knowledge—a sudden general vacuum seemed to have taken over. Despite the success and fame that the movement enjoyed from 1943 on, just a decade later it came to a relatively sharp halt. By 1951 two of its main contributors, von Neumann and Wiener, permanently dropped out from the Macy conferences. By 1953, the last Macy conference took place. Not even informal gatherings ever happened again in America. Whatever happened between 1943 and 1953 that took the world by storm, way beyond the exclusive realms of academic and scientific circles and into the educated average population at large, was not there anymore.

Their British counterpart, the Ratio Club, continued to meet sporadically, gradually dying out during a span of five years, until 1958. By the time the penultimate Ratio Club reunion was taking place in 1955, the American side of the cybernetic movement was practically dissolved.

By that time, having morphed into another project, which did not have the flare, allure, tenets and members of the original project, cybernetics ended up in a relatively obscure program in an American university for one more struggling decade.[53] Aside from this attempt to keep it alive in a substantially different form, the ground shaking ideas firing up bold statements in the media and elsewhere were all but gone. For all intents

---

[53] This "post-cybernetic" approach started when the Ratio Club officially ended, in 1958, at the Biological Computer Laboratory at the University of Illinois in Urbana Campaign—mostly funded by the military. This follow-up movement went by the name of "second-order cybernetics" (or "cybernetics of cybernetics"). It was led by Heinz von Foerster, the physicist who was given the task of recording the proceedings of the Macy Conference, and it housed the cyberneticist Ross Ashby. Due to lack of funding, and to von Foerster's retirement, it shut down in 1970. It always stood as a relatively obscure enterprise without any of the flavor or cache that characterized cybernetics two decades before—only known to hardcore "neo-cybernetic" followers. For an account of the fate of this laboratory, see Müller & Müller 2007.

and purposes classical cybernetics was effectively dissolved and its electrify-ing vibe dead by 1954.

This is an astonishing development, even today, after the fact. Six decades later, we are still asking ourselves what happened. How can one explain this? How can a movement that merited the appraisals (sometimes tinged with fear) of thinkers such as Heidegger and Lacan, that was touted as the science of the future by eminent thinkers, that received substantial funding by the state (at least on the American side),[54] and that was hailed by media as the next step for humanity, come to such an abrupt and unremarkable end?

Several attempts at answering this inquiry have emerged—each tackling one or more levels of explanation. In fact, there have been attempts at articulating this demise coming from several regions of knowledge. Each one has its own strength, and each one points to different aspects of the troubled project toward its end. Some bring up economic reasons for cybernetics' death. Competing cognitive proposals soaked up all the avail-able funding for this type of research, effectively leaving cybernetics to dry out. Some others point to the inherent tensions alluded to above. Some even pin it to reasons of a personal nature. Without dismissing reasons of funding, uneasy interdisciplinarity, or even conflict of personalities, a philosophical assessment has not been sufficiently laid out so far to seek the reasons for the demise of cybernetics.

Indeed, entering deeper into core dimensions of cybernetic theory one could find enlightening signs pointing to what could have happened, at the level of the solidity of its theoretical framework. Most importantly for this book, these cues revolve around the development and evolution of the nature of a machine. This evolving notion was embedded within the cybernetic mindset deeper than many cyberneticians probably realized. Plunging deep into the investigation of the nature of a machine does provide angles to understand better whatever happened to cybernetics.

---

[54] British cyberneticians pursued their endeavor more in a hobby-like style, putting their ideas to work in their free time. During the war, funding was funneled to research on computation—having as a prime aim the decryption of the enemies' communication (as opposed to American research on computation for building tables for anti-aircraft weaponry, and eventually, the atomic bomb). After the war, the rebuilding of Great Britain was the main target of funding. Government funding was never officially provided to the British side of cybernetics.

As well, it opens up doors for understanding later and current approaches to science that carry the cybernetic signature.[55]

But first, I will provide an updated overview of the typical reasons so far adduced for the death of cybernetics.

TRADITIONAL EXPLANATIONS FOR THE COLLAPSE OF CYBERNETICS

Perhaps the most well-known version attempting to explain cybernetics' demise is that it simply ran out of funding. This story emphasizes the alleged fierce competition that streamed from some of the people that were ironically under the influence of senior cyberneticians, but that later had antagonistic proposals of their own. The story line links cybernetics to artificial neural networks in an essential fashion—courtesy of having both McCulloch and Pitts' networks as a common ancestor. The gist of this explanation relies on the claimed fact that a competing "symbolic architecture" approach was shown to be more capable of dealing with some problems, such as theorem proving and chess playing. This dominant change of perspective, coming from the founders of what became later known as Artificial Intelligence (AI), supposedly funneled to itself most of the available funding. This money—still mainly being granted by military offices—went to fund the symbolic approach, leaving cybernetics out in the cold.

Marvin Minsky and Seymour Papert, both disciples of McCulloch, published a book severely criticizing Frank Rosenblatt's "perceptron"[56] —a core element of neural networks, and purportedly, by extension, a core element of cybernetics. It supposedly showed the inherent limitations of single neurons in sophisticated thought processes, a weakness particularly shown in the area of problem solving. The implication was that if the artificial neuron was not capable in small realms (lacking as a standalone or in simple networks), then it could not be good in the big ones (more complex networks would never be able to develop higher processes). This report[57] was regarded by many as the final nail in the

---

[55] More on this on Chapter 10.
[56] Rosenblatt 1958.
[57] Minsky and Papert 1969.

coffin for network-related research—and *a fortiori*, for the distancing from cybernetics as an acknowledged influence.[58]

At least two main criticisms can be launched against such an account. Firstly, as it has been already advanced (and will be further discussed later), although neural networks were part of the core of cybernetics (as we saw above, McCulloch and Pitts' article was indeed a foundational one), it is clear that the networks' approach was not the only one that cybernetics had at the foundation of its own theory. In fact, criticisms against neural networks were launched by von Neumann within the very heart of the cybernetic enterprise, advocating for an alternative view, probably even more cybernetic in spirit—namely, the possibility of building self-reproducing machines. One could even safely assume that Wiener's lack of reaction to von Neumann's severe criticism of neural networks might be explained by the fact that for Wiener the core acceptance of networks was not *a sine qua non* condition for the cybernetic endeavor to continue and flourish. For him there were other more important elements whose widespread acknowledgment was more essential for the cybernetic insight, such as the nature of the message, information, and control.[59]

The other criticism that can be pointed out against this typical account of cybernetics' demise is more historical in nature. By the time that Minsky was criticizing the "perceptron" (1969), cybernetics was largely already dead. This traditional account is thus best located within the context of the origins of cognitive science, particularly the area concerning early neural networks—not within cybernetics at large. And in any case, even if what little remained of cybernetics at the Biological Computer

---

[58] This approach put emphasis on symbol manipulation. On the one hand, it begun to heavily rely on the advance of computers—instead of neurology—and on the other it proposed a systematic structure of mental states, where these symbols were manipulated. It later developed into both Artificial Intelligence and the Computational Theory of the Mind underpinning it. Eventually it was philosophically sanctioned first by Hilary Putnam and Noam Chomsky, and later by Jerry Fodor—who self-admittedly was very hostile to artificial neural networks as a general model of cognition. For an account of this traditional view, both regarding the advantages of the perceptron and the aftermath for connectionism after its criticism, see McCorduck (2004, pp. 102ss); Dupuy (2000, pp. 102ss); Boden (2006, pp. 903ss) and Johnston (2008, pp. 287ss).

[59] See Wiener 1948a, Introduction. Also Conway and Siegelman 2005, ch. 6.

Laboratory at the University of Illinois actually competed for funding with the new and eager generation of AI researchers, it would be inaccurate to affirm that what such "neo-cybernetic" group was asking, namely, funding, equated to cybernetics competing for funding.

In fact, there is the view that the very existence of the Biological Computer Laboratory was the living proof that cybernetics was officially dead.[60] The so-called "Second-order cybernetics"[61] attempted to include the observer into the system, dissolving the whole mechanizing cybernetic view of control. It effectively nullified the bold proposals being advanced by classical cybernetics a decade before, leaving its strong ontological stance.[62] It switched to an epistemological preoccupation with the observer of the system, partly inspired in quantum mechanics' concern with the pervasive role of the observer in any possible measurement within subatomic phenomena. Von Foerster was indeed a physicist by training and was keenly interested in quantum problems right from the moment he entered the first Macy conference. All these "autopoietic" insights, which brought Humberto Maturana and Francisco Varela to his laboratory (escaping from Pinochet's Chile and infused with "the Buddha and the French phenomenologist, Maurice Merleau-Ponty",[63] as well as with some allegedly Heideggerian insights), could have evolved into something unique and interesting—and maybe it did.[64] But what is clear is that

[60] See Asaro 2010.

[61] Previous section.

[62] For Andrew Pickering, cybernetics was substantially more about ontology than about epistemology, instantiated in the mechanical theater of "monsters" (Ashby's homeostat, Grey's tortoises, Shannon's rats) that it produced. See Pickering 2002.

[63] Dennett 1993.

[64] Cf. Varela et al. 1991. This book, which combined phenomenology, cognitive science and Buddhism, enjoyed a moment of fame within the areas of philosophy of mind and cognitive science itself. Roberto Cordeschi's assessment of this proposal is relevant:

> It is still unclear whether these positions bring any advancement in our understanding of cybernetic-related phenomena. On the other hand, many important and legitimate requirements underlying these positions seem to be already fulfilled by the very tradition that they are challenging, whenever the latter is not caricatured or oversimplified (Cordeschi 2008b, p. 195).

second-order cybernetics resembled little to the original cybernetic project. So, if it is indeed the case that it replaced it, then one could conclude
that cybernetics was, in a very real sense, dead.

What is more, a further detail can shed some light on the misconstruc-
tion of identifying the advance of AI as spelling doom for cybernetics. The
generation of disciples that followed the cybernetic masters went the
extra-mile to deny acknowledgment of having its parental roots squarely
set in cybernetics. After the attack of AI to networks research, anything
related to the latter—including cybernetics—would suffer being side-
stepped for funding. This "guilty by association" context might have
had deeper reasons than just playing safe for obtaining grants. In fact, it
is an area of history of science that has been largely unexplored.[65] Again,
this explanation, situated at the beginning of cognitive science (1960s)
accounts for a time when cybernetics was already dissolved. Denying the
cybernetic origin of cognitive science is an issue worth studying in its own
right, but it gives little in terms of explaining why the theoretical parent
experienced atrophy in the first place.

Still related to theoretical issues, there is another one that has lately
attracted more attention, but that is nevertheless intertwined with issues of
personal nature. It has to do with what occurred after von Neumann's letter to
Wiener, regarding the inadequacy of McCulloch and Pitts' neural networks
for articulating the human mind—and von Neumann's subsequent proposal
for exercising a "back to the basics" move, instantiated in the study of the self-
reproductive capabilities of a virus. This (in)famous missive was sent after von
Neumann delivered the opening talk at the first Macy meeting—and actively
participated in the second one. The letter also set up a meeting with Wiener,
which seemingly took place in December of the same year (1946). Regardless
of what happened in that meeting, set out to discuss the discomfort vented in
the letter, it was evident for many that a rift between the two heretofore close
friends happened—and this rift was sometimes expressed in eccentric ways. It
was reported that Wiener would loudly snore at von Neumann's talks. Von
Neumann, in turn, would loudly flip the pages of the *New York Times*, seated
in the front row, while Wiener was giving lectures.[66]

---

[65] Piccinini's (2002) mentioned above critical review of Dupuy's (2000) book on
cybernetics points out to this absence.

[66] See Heims 1980, p. 208. In all fairness, Wiener's loud snoring is part of the
collection of his eccentricities—mainly due to the fact that once awaken, he would

Whatever the effect the letter might have had, there were reasons for Wiener's reaction—even if delayed. For starters, when the friendship between both men was at its pinnacle, Wiener's lobbying at MIT managed to procure for von Neumann a position at the mathematics department. MIT had a long tradition of engineering, while Princeton had a number of Nobel Laureates. Wiener was likely interested in bringing that kind of flare to the cybernetic hub—which for the lack of an official institutional setting, was de facto occurring at MIT. Von Neumann used this important offer as leverage to persuade Princeton to give him funding for building his own computing machine—which he obtained, promptly refusing MIT's invitation.

As if that maneuver was not enough, von Neumann asked Wiener to "lend" him his chief engineer, Julian Bigelow, whose genius was so crucial for building his predictor—and Wiener agreed! The double irony of this move is that the machine to be constructed had as its main aim the aiding of von Neumann's secret assignment—the building and improvement of atomic bombs. All these while Wiener had already become deeply disturbed on account of the destructive capabilities of nuclear power—and hence the grave moral responsibilities of men of science.

Indeed, von Neumann was an active advocate of a pre-emptive strike against the Soviet super power. In sharp contrast, Wiener was severely affected after the bombing of Hiroshima and Nagasaki; his stance regarding war changed 360 degrees, making him a vociferous peace supporter. In fact, this newly acquired stance, which was represented in his resolute decision to walk away from all government and military contracts, was seen with suspicion by the McCarthyism of the Cold War era. This development has been identified as one more reason why cybernetics fell out of grace. Topping it all off, Wiener kept a close relationship with his Russian peers in science—who had developed a keen interest in cybernetics, and were largely backed by the Soviet state. This might have triggered a negative aura around Wiener and his project—a "cautious" view held by the government.[67]

provide the most penetrating insights regarding what was said during his sleep (1980, pp. 206–207). The difference is that this feature was no longer taken as a picturesque occurrence; rather, von Neumann took offence from it, triggering his conscious noisy behavior while Wiener was lecturing.

[67] Conway and Siegelman 2005, chs. 12 & 13.

As to what strained his relations with von Neumann, we have no proof, but Wiener might have caught wind of von Neumann's side agenda. Or he might have had a delayed reaction regarding von Neumann's Janus-faced "move" to obtain funding for his computer machine at Princeton—at the expense of Wiener's good intentions. Or perhaps it was a combination of the above.

Not all outward resentment came from the side of Wiener, however. It is acknowledged that when Wiener made the link between his cybernetic incursions into control, communication and negative feedback on the one hand, and thermodynamics and chaos on the other (thus identifying information with negative entropy) von Neumann took it personally—and rather badly. Von Neumann had been also studying chaos for years, and he made a name for himself from his earlier years in the realm of the mathematics of quantum mechanics. However, he did not elaborate his insights into chaos and entropy in connection with information to the point that Wiener did. In fact, there might have always been a rivalry particularly coming from von Neumann, who probably felt that Wiener always had the upper hand in mathematics—after all, the latter had been recognized as a veritable genius since his childhood. Von Neumann became jealous that Wiener established the startling correlation and not him. So, what might have started as friendly academic competition could have developed into mutual hostility.[68]

There was still a further detail that might have played a role, above and beyond the distancing with von Neumann—in fact, with Wiener's distancing from the whole cybernetic group: An occurrence that has lately received more attention in terms of its possible influence on the dissolution of the whole enterprise.

Margaret Wiener, who was literally "shipped in" from Germany to marry Norbert, belonged to a severely conservative Protestant Puritan tradition. Also, she was a fairly overt admirer of Adolf Hitler, and her family, proudly Nazi—one of her brothers ran a concentration camp. Barbara, their first daughter, one day got in trouble in elementary school because she candidly talked about the (Nazi) literature her mother avidly read at home. Apparently, the secularism of Wiener's parents might have been sufficiently profound for Margaret to ignore Norbert's Semitic roots.

[68] Ibid., pp. 165–166.

Rather, it was clear that for Margaret, the fame and stature of her husband on the world stage, and hence her social position, was the absolute priority. The whole family, Norbert included, went to a church of her denomination in a disciplined manner.

Since it is until now unclear what happened between both men, it is not out of the realm of possibility that Margaret, at a certain moment, turned her husband's mind against von Neumann—a Jew (despite the fact that his family adopted Catholicism in the 1930s). They both already half-mocked von Neumann's impeccable dressing and ultra-formal manners.[69] She knew how to get Norbert Wiener to steer things her way—as it will be shown below. Margaret's ideas were likely already beginning to be distilled, following her own motto,[70] despising the ways of a liberal life. Von Neumann divorced and remarried after his first wife ran away with a graduate student of his.

If Wiener grew apart from von Neumann, chances are that the latter was not profoundly affected, since von Neumann always kept some distance from the cybernetic group anyway. He was in the group, without being part of it, so to speak. This might be explained, beyond his own way of being—a cosmopolitan man gifted with political maneuverability—with the sort of secret appointments that he took for the most part of his career. But the distance that occurred between Wiener and the rest of the group certainly had dramatic consequences. In 1951, three years after the publication of the book, Norbert Wiener severed all ties with his close friends and cyberneticians, particularly with Warren McCulloch and Walter Pitts. This episode has not attracted enough attention until lately,[71] and a brief exposition of what occurred seems fitting, in order to assess the degree of influence that this episode had in the entire cybernetic endeavor.

From the moment that Wiener and McCulloch had met, a gradual but eventually close friendship was formed; it was only to grow more intense in

---

[69] Ibid., p. 144.

[70] Margaret Wiener wrote down on a piece of paper what would probably summarize her outlook on life:

> One way to arrive at the aristocracy
>   if you aren't born there
>   is to eschew all forms of liberalism (ibid., p. 335).

[71] Ibid., pp. 213–234.

the subsequent years after the publication of their articles in 1943 and during the years of the Macy conferences. McCulloch acquired a ranch (with a farm) in Old Lyme, Connecticut, which he renovated and put to work. This place was a hub for all the intellectuals, scientists, and academicians that were around for a conference, or who just wanted to drop by and mingle with the McCullochs for a while. In fact, the place was big enough for guests to spend the night—even several days.

Reportedly, Norbert Wiener always was in the happiest of moods while he was at the ranch. The lake, the nature, the late chats over drinks over entire weekends, all these were a relaxing antidote for Wiener's frequently stressed mind. The McCulloch's were a remarkably liberal couple and such things as swimming at the lake in the nude was customary—for them and their guests. At least one person recalls the usual sight of Norbert Wiener on the lake, huge belly up, cigar in hand, loudly talking while floating away. Such was the happy, if permissive, ambience at the McCulloch's ranch.

Needless to say, Margaret Wiener would have never approved of this conduct. In fact, unlike the other wives, she never accompanied Norbert to any of these weekend getaways. Warren McCulloch's pomposity, eccentricity, and for those times, markedly liberal ways, were more than what a conservative Puritan German could take—and the open repulsion against Wiener's friend was immediate from the first time they met. Wiener's daughters, Barbara and Peggy, recall that with absolute certainty, their mother Margaret never found out about this "debauchery". Her reaction would have been, in all probability, nothing less than cataclysmic. Margaret Wiener's disgust for this circle of friends triggered a long-term plan that seemed to have worked out at the end.

In 1951, three years after the publication of *Cybernetics*, and having published another successful book,[72] Wiener was trying to publish his memoirs—to no avail. Publishers found the manuscript to be too vitriolic and ridiculing of some people who had been Wiener's mentors. Even Wiener's portrayal of his own self was embarrassing. The intended book was coming across as a personal rant against both the author himself and people who were still alive. Publishing companies did not find the material

---

[72] *The Human Use of Human Beings* (Wiener 1950), unlike *Cybernetics* (Wiener 1948a), was intended to serve as an introduction to the new science for the layman. As previously mentioned, the 1948 book became a best-seller against all odds.

attractive. Some found it even parochial.[73] The uniform rejection, even from the publishing house at his own home institution which previously fought for co-publishing *Cybernetics* (MIT), was taking a toll on Wiener's state of mind. In fact, it was exacerbating the intense depression into which he submerged himself out of writing about his own (not particularly happy) childhood. Wiener became as vulnerable as a seriously depressed person can get. His wife Margaret saw the opportunity to advance her plan of secession.

Some years before, when Barbara had begun her undergraduate studies in Chicago, Warren McCulloch expectably offered his house for her to stay, since it was close to her school. Wiener was grateful for the offer, accepted it, and sent Barbara to stay there for quite some time. The McCullochs were so attentive about her staying with them, that once they even confronted her regarding a medical student she began dating. She was 19 years old and they felt responsible. It was all innocent and nothing came out of it, but history turned that episode into an irony.

Recall that this was the same period during which McCulloch gave shelter to Pitts and Lettvin. Margaret Wiener told Norbert, at the lowest point of his depression, that during the time when Barbara was staying with the McCullochs, "the boys" (as the young cyberneticists used to be called) seduced their young daughter more than once. And more than one.

Norbert Wiener went into shock. And rage ensued.

Margaret Wiener had an extreme fixation with the perverse (for her) nature of sex, having submitted her daughter Barbara to several humiliations—all sex-related, and all implying she did something gravely inappropriate, when she did not. The Puritanical rattle was sometimes so evident, that even Norbert Wiener, despite his legendary absent-mindedness,[74] on more than one occasion came to the rescue of his daughter from the intrigues of the mother.[75] However, by this time, Wiener's psyche was at an all-time low,[76] allowing Margaret to instill venom in his defenseless mind. Given the (sexual) nature of the

---

[73] As a publisher put it while rejecting the manuscript, "a book of almost wholly American interest." (Conway and Siegelman 2005, p. 218).

[74] Jackson 1972.

[75] Conway and Siegelman 2005, pp. 199–200.

[76] A psychologist friend of Wiener recounted how he would cry uncontrollably when he would begin to talk about his past—even while in the midst of a meal with him at a restaurant (Ibid., p. 218).

accusation, and given the social ways of the time, Norbert was never going to further stir things and damage his daughter's reputation by inquiring about the truth of this revelation—outcome that Margaret likely knew quite well, successfully sealing her master plan.

Around the same time, Walter Pitts and Jerome Lettvin were in the best of moods. The Radiation Laboratory (Rad Lab) mentioned in the previous chapter was morphed into the Research Laboratory of Electronics after the war ended. This new lab had more than sufficient funds for serious research, and had Warren McCulloch as its head— for which he left his permanent professorship at the University of Illinois in Chicago. As if that was not good enough, the laboratory was going to have a research bent toward human cognition inserted in the "bigger picture" of military research and development. Pitts and Lettvin, with their new fancy machines and bright prospects of adventurous investigation, were ecstatic and overjoyed. In accordance with their mood, they wrote a bombastic letter to Wiener (and Rosenblueth), the language of which was supposed to be taken as a prank. It started with the words "Know, o most noble, magnanimous and puissant lords..." In all likelihood, Wiener would have found it amusing, had he found himself in a "normal" state of mind. Margaret's "revelations" of his daughter's permanent "stain" at the hands of these "boys" had occurred just the night before. Wiener's answer was terse in the extreme. The telegram that he sent in response to a colleague of theirs, so that he passes on the message, read (caps original):

IMPERTINENT LETTER RECEIVED FROM PITTS AND LETVIN. PLEASE INFORM THEM THAT ALL CONNECTION BETWEEN ME AND YOUR PROJECTS IS PERMANENTLY ABOLISHED. THEY ARE YOUR PROBLEM. WIENER.[77]

As for McCulloch, Wiener opted instead for applying a permanent "silent treatment", while at the same time vituperating his persona, academically and professionally, among common colleagues. Both McCulloch and Pitts (and Lettvin) were first in shock. Then in denial. For a while, they thought

[77] Ibid., p. 219.

that it was going to be a passing phase,[78] that things would later come back to normal, and that they would find out what happened. They simply could not, and did not, understand what took place. Gradually, however, it became clear that Wiener's decision had a permanent character—and a dark cloud settled in for everyone.

McCulloch took it badly. Although he would continue with a "business as usual" attitude in his life and work, it was clear for those surrounding him that he was hurting and merely "pushing through"—surviving. The years of joyous friendship, of meeting for the most fulfilling conferences and gatherings, of seeing eye-to-eye on fundamental ideas about the world . . . these were not going to go away easily. Wiener himself confessed later that the break up with his colleagues and friends was taking a toll on him, not only mentally, but also psychosomatically: he alluded to having contracted a painful heart condition due to this episode. By all accounts, however, the one who took it the worst, and by far, was Walter Pitts.

Pitts regarded Norbert Wiener as the father he never had. Even if McCulloch exercised a de facto "adopting" of him by letting him stay at his house when he was homeless, McCulloch did the same for several other students. McCulloch was, to put it simply, just a generous man who cared in a very concrete sense for his students' well-being. With Wiener, however, it was different. Wiener brought Pitts to MIT, vouched for him, put him under his direct supervision, and lobbied for scholarships, grants and job positions for him. Wiener only had two daughters—as smart and intelligent as they were, given the times, he might had desired to have his legacy passed through to a son. In pressing Pitts to perform his duties and accomplish his goals (Pitts had a severe habit of procrastination), Wiener was very likely re-enacting the rigor that his own father had with him for his own good. Pitts, who escaped an uneducated environment where he used to be beaten, found in a very literal way, a father.

After Wiener dispatched the above missive (and knowing that Wiener was mouth trashing him and McCulloch), Pitts entered into a spiral of self-destruction. He burnt all the research and writing that was going to grant him a Ph.D. on a three-dimensional model of neural networks—as

---

[78] Wiener had short episodes of crisis in the past, overwhelmed by work and responsibilities. James Killian Jr., president of MIT at the time, recalled having in his desk drawer a folder with all the letters of resignation Wiener handed over the years (Ibid., pp. 220–223).

mentioned before, a proposal that baffled with admiration mathematicians at the time. He isolated himself from everyone. He stopped talking. He began to drink heavily. And he remained in that deplorable state until his death, which was alcohol-related (cirrhosis).[79] Revealingly, Pitts died only five years after Wiener, despite being considerably younger than him.[80]

Pitts' friend, Jerome Lettvin, the last cybernetician alive, died in 2011. Later in his life, he was asked to give an account of Norbert Wiener at a conference in Genoa, which he did with most empathy and in the best possible manner. Six decades later, at almost 80 years old, Lettvin was still hurting from the break up. Ten years after the unfortunate occurrence, Arturo Rosenblueth finally told him what happened, recounting the machinations that Margaret Wiener put in place to secure their separation.[81] In spite of it all, Lettvin respected and admired Wiener until the end. Wiener's widow, Margaret, was at the same conference, and approached Lettvin to thank him for his kind words about her late husband. This is how Lettvin recalls the encounter:

> I prepared a very careful, very adulatory talk and, afterwards, Mrs. Wiener came up to congratulate me and offered her hand—you know, she was a slight woman—but I really wanted to hit her as hard as I could, because I knew that she had contrived the break.[82]

Jerome Lettvin did not want to open up much regarding this painful episode for him and his friends, but being the last cybernetician alive, he did set the record straight before his recent death in 2011. These

---

[79] See Smalheiser 2000.

[80] It may be the fair to say that the story of Walter Pitts is until now one of those great untold stories in the history of science. It might be the lack of sufficient written records about him, or that he did not *formally* accomplished much himself. But great minds around him reported the reliance of their own upon his help. He remains a sort of myth surrounding cybernetics: The man whose mind baffled Norbert Wiener (a genius himself), Bertrand Russell and Rudolf Carnap, and who ended up in total obscurity (See Easterling 2001).

[81] Rosenblueth was present when Margaret disclosed such "revelation" to Norbert, during dinner. He kept it for himself for several years (Conway and Siegelman 2005, pp. 224–225).

[82] Ibid., p. 223.

revelations have spanned a new round of possible explanations for cybernetics' decline. Given the powerful effects that passion and human emotions may have on historical occurrences of long lasting consequences, one could be tempted to reduce the whole development of cybernetics' implosion to a unique, selfish, and hidden act of a sole person. After all, Henry VIII's infatuated lust for Anne Boleyn in the seventeenth century ended up in the subsequent breakup of the entire UK from Rome, effectively creating an unsurmountable wedge in a divided West.[83] Could it be that one person's "matters of the heart" affect in such a profound way the future of an entire worldview? Indeed Oliver Selfridge[84] said of this gossip that

> It really fucked up cybernetics...because here you've got the guy who invented the term and invented the idea right there with you, but there was no interaction at all with Norbert, which was a crying shame.[85]

Without attempting to subtract the causal power from the passionate act of a woman with twisted intentions upon a man's overreaching actions (be it Anne Boleyn or Margaret Wiener), one could, however, recall the traditional advice given to any first-year philosophy student: A cause for

---

[83] In a paragraph brilliant in its succinctness, Elizabeth Eisenstein summarized the dramatic split of the West thus:

> Sixteenth-century heresy and schism shattered Christendom so completely that even after religious warfare had ended, ecumenical movements led by men of good will could not put all the pieces together again. Not only were there too many splinter groups, separatists, and independent sects who regarded a central church government as incompatible with true faith but also the main lines of cleavage had been extended across continents and carried overseas along with Bibles and breviaries. Within a few generations, the gap between Protestant and Catholic had widened sufficiently to give rise to contrasting literary cultures and lifestyles. Long after Christian theology had ceased to provoke wars, Americans as well as Europeans were separated from each other by invisible barriers that are still with us today (Eisenstein 2012, pp. 172–173).

[84] One of the latest and youngest additions to the cybernetics group before the break up, he died in 2002.

[85] Conway and Siegelman 2005, p. 233.

an effect is seldom exhaustive. One particular situation may have several valid explanations intertwined, spanning across several levels of discourse and reality. In this section, some of these levels have been laid out— theoretical, social, economic, political, and personal. But projects that advance bold ontological (and epistemological) proposals in a robust way tend to supersede their own founders. In the words of Wiener himself,

> We have contributed to the initiation of a new science which . . . embraces technical developments with great possibilities for good and evil. They belong to the age, and the most any of us can do by suppression is to put the development of the subject into the hands of the most irresponsible and the most venal of our engineers.[86]

It is reasonable to entertain the possibility that there is space for a further philosophical interpretation of what happened with cybernetics at the theoretical level. This is particularly appropriate given the fact that after the breakup, some cyberneticians were not only still active with a classical cybernetic outlook on things, but indeed took the project to the next level in different ways—indeed away from "irresponsible and venal engineers". This was precisely the case with William Ross Ashby and John von Neumann. Identifying what went on in each case will, I hope, further contribute toward the understanding of a notion of machine, and with that, shed some light on what possibly occurred with this movement. But first, I will lay out in the next chapter the context that allowed for the rise of cybernetics in the first place.

[86] Wiener 1948a, p. 28.

CHAPTER 4

Pre-Cybernetic Context: An Early Twentieth-Century Ontological Displacement of the Machine

THE "FOUNDATIONAL CRISIS OF MATHEMATICS" AND THE RESPONSE FROM FORMALISM

In the second half of the nineteenth century, the German mathematician George Cantor (1845–1918) engaged in a series of investigations regarding infinite numbers. Some of his observations were considered disturbing by some schools of thought in mathematics, and even more so the conclusions that were being reached from them. A typical Cantorian reasoning would lead to a paradox involving operations with transfinite numbers. For instance, if we have an infinite set of natural numbers, and a set conformed by all the sub-sets of natural numbers, which set would be larger? The notion of a cardinal number is introduced to account for the situation in which two infinite sets would still give an infinite set. This seemed self-contradictory and the tension was denounced, since it was occurring at the foundation of arithmetic—at the very core of mathematics.

The association of the notion of the "infinite" with the idea of God, put in place since early Medieval philosophy, was still present in its own nuanced way, at the end of the nineteenth century. Indeed, Cantor thought of himself as being the receptor of the mathematical

© The Author(s) 2017
A. Malapi-Nelson, *The Nature of the Machine and the Collapse of Cybernetics*, Palgrave Studies in the Future of Humanity and its Successors, DOI 10.1007/978-3-319-54517-2_4

insight regarding infinite numbers as a message delivered by God. In an occurrence that shows how controversial were his investigations at the time, Leopold Kronecker, Cantor's own mentor, reportedly stated that "God made the natural numbers; all else is the work of man".[1] Adding these metaphysical underpinnings to the extent of the consequences of his musings, might help clarify the reasons behind the subsequent furor against Cantor—who eventually broke down psychologically, ending his days in a mental ward.

This whole episode is a prelude to what was known as the "Foundational Crisis of Mathematics" (*Grundlagenkrise der Mathematik*). The four main schools of thought in mathematics (logicism, formalism, intuitionism, and Platonism) reacted to the denounced paradoxes. There is widespread consensus that the first two, logicism and formalism, which were substantially related, got the upper hand in responding to the crisis. Each displayed its own particular emphasis on the treatment of the problematic situation, and each triggered a different (although again, related) effect on mathematics, philosophy, and science.

The logicist attempt at giving a solid foundation to mathematics resulted in efforts to reduce it to an extension of logic. But this route had some pre-conditional issues to be resolved. Immanuel Kant famously stated that analytical judgments do not advance knowledge. Following the principle of non-contradiction, analytical propositions merely contemplate their own concepts as their object of understanding, finding the idea that was already embedded within them (e.g., a triangle has 180 degrees). Logic is the analytic discipline *par excellence*: obeying the law of non-contradiction, it engages in the contemplation of itself as an object of study. Although it provides no augmentation of knowledge, its own nature made it remain supremely stable, not having changed since the time of Aristotle:

> For the advantage that has made it so successful logic has solely its own limitation to thank, since it is thereby justified in abstracting—is indeed obliged to abstract—from all objects of cognition and all the distinctions

---

[1] The quote in German uses the word "integers" instead of "natural numbers"; however, the latter is the way in which the quote is traditionally remembered (Bell 1937, p. 527).

between them; and in logic, therefore, the understanding has to do with nothing further than itself and its own form.[2]

Thus, the rescuing strategy for attaining a solid body of mathematics would somehow have to base mathematical knowledge on the secure foundations of logic. However, the mere mentioning of the notion of mathematical *knowledge*, which implies an epistemic gain, already suggests that we are facing a stumbling block—since there is no augmentation of knowledge from strictly analytical grounds. However, mathematics, by most accounts, indeed provides an increase of knowledge. And this goes against, to repeat, the very nature of an analytical enterprise—such as logic. After all, logic is secure, but at the price of not having changed in millennia; whereas mathematics has clearly advanced. In fact, Kant thought of mathematics (specifically of arithmetic, geometry, and algebra) as being synthetic *a priori*—just as the judgments that form the natural sciences (e.g., physics) are, where the advances after Newton were staggering. Kant defended arithmetic as synthetic in one of the passages that has garnered perhaps the most dissatisfaction among his readers, ever...

We might, indeed at first suppose that the proposition 7 + 5 = 12 is a merely analytical proposition, following (according to the principle of contradiction) from the conception of a sum of seven and five. But if we regard it more narrowly, we find that our conception of the sum of seven and five contains nothing more than the uniting of both sums into one, whereby it cannot at all be cogitated what this single number is which embraces both... We must go beyond these conceptions, and have recourse to an intuition which corresponds to one of the two—our five fingers, for example... For I first take the number 7, and, for the conception of 5 calling in the aid of the fingers of my hand as objects of intuition, I add the units, which I before took together to make up the number 5, gradually now by means of the material image my hand, to the number 7, and by this process, I at length see the number 12 arise. That 7 should be added to 5, I have certainly cogitated in my conception of a sum = 7 + 5, but not that this sum was equal to 12. Arithmetical propositions are therefore always synthetical, of which we may become more clearly convinced by trying large numbers. For it will thus become quite evident that, turn and twist our conceptions as we may, it is impossible, without having recourse to

[2] Kant 1787, B ix.

intuition, to arrive at the sum total or product by means of the mere analysis of our conceptions.[3]

However, the German mathematician Gottlob Frege (1848–1925) took upon himself the task of showing that arithmetic, which does provide augmentation of knowledge, is in fact an analytic discipline. The question is: How to maintain the solidity of logic without removing the capability of furnishing new knowledge? In his *Foundations of Arithmetic*, he severely criticized the idea instantiated in the paragraph above—which to Frege's credit, did trigger much discomfort not long after Kant provided such "visual" explanation. Frege's attack, to some degree commonsensical, involved the multiplication of very large numbers, the premise being that one does not refer to spatial objects (e.g., fingers) when operating with those...

> [I]s it really self-evident that 135,664+37,863=173527? It is not; and Kant actually urges this as an argument for holding these propositions to be synthetic... Kant thinks he can call on our intuition of fingers or points for support, thus running the risk of making these propositions appear to be empirical, contrary to his own expressed opinion... [H]ave we, in fact, an intuition of 135,664 fingers or points at all? If we had, and if we had another of 37863 fingers and a third of 173527 fingers, then the correctness of our formula, if it were unprovable, would have to be evident right away, at least as applying to fingers; but it is not. Kant obviously was thinking of small numbers. So that for large numbers the formula would be provable, though for small numbers they are immediately self-evident through intuition.[4]

Having accused Kant of underestimating the power of analysis (in the operationally fruitful sense), the ground was ripe for declaring that arithmetic is *both* a knowledge-augmenting discipline *and* an analytic practice. What was being accomplished was no small feat, as Frege was keenly aware. He was one small step closer to the ambitious Leibnitzian dream of subsuming everything under heaven to the power of *calculus rationalis*, getting thus nearer to the sort of idealized *visio Dei* always longed for by philosophers. Now that arithmetic could be spoken of as analytical

---

[3] Ibid., B 205.
[4] Frege 1884, p. 6.

(without destroying its capacity for knowledge-rendering), the path was clear for mapping it upon that other discipline known by its secure foundations: Logic. Being analytical no longer was a cardinal sin against the augmentation of knowledge. *A fortiori*, a reduction of mathematics to logic seemed now as a course of action worth pursuing.

As indicated above, there was more than one response to the foundational crisis that mathematics was experiencing. Besides the logicist account outlined before, what came to be known as the formalist approach exercised an impact in the history of mathematics that spilled over to neighboring areas of possible research, which gradually morphed into fields of their own. The German mathematician David Hilbert (1862–1943) envisioned the construction of the whole mathematical edifice not based upon logic—as Frege, Whitehead, and Russell would like. He rather preferred anchoring mathematics upon foundational conventions to which mathematicians would arrive after sorting out some basic issues that were marring, according to Hilbert, the solid web of deductive truisms characteristic of mathematical knowledge.

Accordingly, at the International Congress of Mathematicians of 1900, convocated in Paris to celebrate the turn of the century, Hilbert addressed the audience in an original manner. Instead of delivering a presentation that would contribute to the increase of mathematical knowledge, he produced a set of 23 mathematical problems that according to him would set the pace and provide the motivation for the development of mathematics in the following century. Only 10 of these problems were actually presented at the conference—the rest were made available later. Further clarification in the articulation of these problems evolved during the following three decades, and two of them (problems 2 and 10) begot what came later to be known as the "Decision problem". Originally, the 2nd problem (named "The compatibility of the arithmetical axioms"), which aimed at finding out whether the axioms of arithmetic were consistent, was stated in the form of a question, asking...

> *Whether, in any way, certain statements of single axioms depend upon one another, and whether the axioms may not therefore contain certain parts in common, which must be isolated if one wishes to arrive at a system of axioms that shall be altogether independent of one another.*[5]

[5] Hilbert 1902, p. 447 (Italics original).

The 10th problem (entitled "Determination of the solvability of a Diophantine equation")[6] asked for a "procedure" that would allow for the discovery of whether an equation is true or false—at the time, Hilbert circumscribed the task only to "Diophantine equations":

> Given a diophantine equation with any number of unknown quantities and with rational integral numerical coefficients: *To devise a process according to which it can be determined by a finite number of operations whether the equation is solvable in rational integers.*[7]

In 1928 at Bologna, in another meeting of the Internal Congress of Mathematicians, Hilbert expanded the reach of both problems into a more general and relevant set of questions affecting the whole of mathematics. Although Hilbert presented them as open questions, he was of the idea that the answer to each one of them would be in the positive—this effort constituted what came to be known as "Hilbert's Program". These questions, each under the banner of their identifying issue, were thus:

1) Completeness: Are the true propositions of mathematics all provable?
2) Consistency: Is the set of true propositions in mathematics free of contradiction?
3) Decidability: Is there a procedure by which every true proposition could be provable?[8]

---

[6] A Diophantine equation is an equation whose solution can only allow for numbers without fractions—integers, often called "natural" numbers. Recall Kronecker's remark, above.

[7] Ibid., p. 458 (Italics original).

[8] The text of the conference has been translated as "Problems of the grounding of mathematics" (Mancosu 1998, pp. 266–273). Stephen Hawking's somewhat canonical abbreviation of the triad problem reads like this:

1. To prove that all true mathematical statements could be proven, that is, the *completeness* of mathematics.
2. To prove that only true mathematical statements could be proven, that is, the *consistency* of mathematics,
3. To prove the decidability of mathematics, that is, the existence of a *decision procedure* to decide the truth or falsity of any given mathematical proposition (Hawking 2005, p. 1121).

The Austrian mathematician Kurt Gödel (1906–1978) was in all likelihood present at this 1928 talk, which also served as a retirement speech for Hilbert. They never met in person, but two days later, at the same conference, during a round table, Gödel gave a presentation regarding certain aspects of the *Principia Mathematica*. By this time, Gödel had already reached his "First Incompleteness Theorem". The threat to Hilbert's goal was noticed by the previously mentioned mathematician John von Neumann, who took him aside to talk. The next year von Neumann communicated to Gödel that he followed up on his presentation and reached a second incompleteness theorem. By this time, Gödel had already done so as well. In 1931, only one year after the completion of his Doctorate, at 25 years old, Kurt Gödel published "On Formally Undecidable Propositions of Principia Mathematica and Related Systems I".[9] After this paper, quick and wide (but not absolute) consensus was reached in asserting that—at least in the way that Gödel framed Hilbert's first and second new problems (completeness and consistency)—the answer was in fact in the negative. Or rather, that the very idea of wanting a solid system, that is both complete and consistent, was ill formed.

Gödel's "first theorem" asserted that if in a sufficiently robust arithmetic system the set of axioms is consistent, then there will be at least one true proposition that cannot be proved.[10] The "second theorem" showed

---

[9] Gödel 1931. There was supposed to be a follow up paper by Gödel, which never occurred—hence the "I" at the end of the title.

[10] Kurt Gödel's article is famous not only due to his mathematical rigor (accessible only to the trained mathematician), but also due to the fact that its translation into English is an issue in itself. One could safely state that he never was fully satisfied with any of the several translations that took place throughout the years while he was still alive. In a translation by Elliott Mendelson, the one which he was the least unhappy with, Gödel's very technical text excerpt regarding what came to be known as his "first theorem"—in fact, the sixth theorem in the paper—reads:

> Theorem VI: For every ω-consistent recursive class $\varkappa$ of FORMULAS, there exists a recursive CLASS EXPRESSION-r, such that neither v Gen nor Neg (v Gen r) belongs to Flg ($\varkappa$) (where v is the FREE VARIABLE of r) (Gödel 1934, p. 24).

that if a system is consistent, such consistency cannot be proved from within the system.[11] Arithmetic, and by extension mathematics, if understood as founded upon a strong body of propositions, is incomplete if it is found to be consistent: There will be at least one axiom that, even if true, will not be provable from within the system.

These findings were devastating for Hilbert—who never formally replied—and by extension to Russell and Whitehead—whose logicist approach was partly shared by Hilbert for number theory, but was later rejected. In fact, at a certain point, Hilbert *was* in principle amenable to relying on *Principia Mathematica* to formalize arithmetic, since a mechanized way of proving the whole mathematical edifice would be close to his goal of making mathematics an orderly and tightly fit structure. The effect of the incompleteness theorems for mathematics in what regards its

In layman terms, this formulation could be paraphrased as:

> the first theorem...asserts for any formal theory...rich enough to include all the formulas of formalized elementary number theory...that if it is consistent...then it is incomplete (Stoll 1979, p. 446).

By all the formulas of elementary number theory we can understand the whole of arithmetic. It follows that:

> the entirely natural idea that we can give a complete theory of basic arithmetic with a tidy set of axioms is wrong (Smith 2007, p. 13).

It would thus seem that if the set of all arithmetic propositions is complete, then it is inconsistent—and if consistent, then it is incomplete.

[11] The surprisingly clear textual reference in Gödel's paper appears as follows:

> Theorem XI. Let $\varkappa$ be any recursive consistent class of FORMULAS; then the SENTENCE which asserts that $\varkappa$ is consistent is not $\varkappa$-PROVABLE; in particular, the consistency of P is unprovable in P, assuming that P is consistent (in the contrary case, of course, every statement is provable) (Gödel 1931, p. 36).

Here Gödel would deduce a truism that follows from the previous (so-called "first") theorem. If there is a proposition that would prove the consistency of a set of propositions, then this proposition would lie outside the system of such set.

decidability was certainly foreseen by Gödel. However, as we will see, Hilbert's third problem took on a life of its own.

A MACHINAL[12] UNDERSTANDING OF AN ALGORITHM AND THE MATERIAL LIBERATION OF THE MACHINE

The Cambridge mathematician Max Newman had a Master's student in his 1935 class, recipient of a fellowship for King's College. A young Alan Turing (1912–1954), after completing a dissertation in Gaussian theory for his undergraduate degree, joined professor Newman's course, which dealt with fundamental mathematical problems. Noticing that the "Decision problem" was so far deemed unsolved, Turing embarked on its solution. Professor Newman had in his lectures identified the quest for an "effective procedure" for deciding whether or not any mathematical equation is true. What an "effective procedure" stands for would show to be an issue in itself soon.

The notion of "mechanical" was not foreign in mathematical logic, particularly within logicism. An entity "affecting" something implies a certain action, a change. A procedure that affects a change in a deterministic manner could be spoken of as being mechanical. Alan Turing was receptive to this qualification. Further equalizing the notion of a mechanical procedure to that of computability, he aimed to show in an essay that not the whole of mathematics is in fact computable—thus answering Gödel's third problem (whether there is an effective procedure to prove any true mathematical proposition) in the negative.

---

[12] This archaic English word is used at times when translating the work of some Continental thinkers (e.g., Heidegger, Derrida), particularly when these refer to an aspect of the entity "machine" less related to the typical feature by which it is usually known—its functionality, or the processes it underpins—and more related to its nature as such—a machine *qua* machine (whatever it is that makes machine to be what it is, independently of the *mechanical* processes it supports). Although an inquiry into the nature of an algorithm, as we will see in this chapter, followed the said typical version (the machine-like behavior of an algorithm subsuming the latter as a type of machine), it will later become evident that it opened up a broader understanding of what a machine stands for. In one of its possible meanings, the derivative "machination" stems from this aspect of the notion of a machine. More on this in Chapter 5, Section "Pinning Down the Core of Cybernetics" of this work.

Just before publishing the paper, Turing (and Newman) found out that both Gödel and the American logician Alonzo Church (1903–1995) were separately working on the decision problem at roughly the same time. In fact, Church had already published his findings in 1935. Expectably, Newman was at first reticent in supporting Turing publishing his paper, but later he was convinced that the angle was sufficiently different, giving Turing the get go—which happened in early 1936. The name of this historic paper was "On Computable Numbers, with an Application to the Entscheidungsproblem".[13]

Turing took the "mechanical procedure" seriously enough as to devise an abstract machine that would function *qua* algorithm—the very definition of which being sought after by Gödel (via the notion of recursive function) and Church (by means of his own mathematical construction, the "lambda calculus"). Key notions in Turing's imaginary engine were determinism and the "brute" following of instructions. The Turing machine (as it quickly began to be called) was constituted by an endless tape, a reading "head" (which can also write and erase) and a set of instructions. It works in an ingeniously simple way. The endless tape is divided into squares (just as the old film tapes), and each square can have written the symbols 0 or 1 in it; or it can be empty. The instructions will be constituted by a set of commands that will instruct the head what to do if it finds, say, a "1" (erase, write "0", move to the right) or a "0" (erase, move to the left) or an empty square (write "0", move to the right), or another "1" (move to the left), and so on.[14]

---

[13] Turing 1937

[14] This is how Turing described it:

> The machine is supplied with a "tape" (the analogue of paper) running through it, and divided into sections (called "squares") each capable of bearing a "symbol"...We may call this square the "scanned square". The symbol on the scanned square may be called the "scanned symbol". The "scanned symbol" is the only one of which the machine is, so to speak, "directly aware"...In some of the configurations in which the scanned square is blank (i.e. bears no symbol) the machine writes down a new symbol on the scanned square: in other configurations it erases the scanned symbol. The machine may also change the square which is being scanned, but only by shifting it one place to right or left (Turing 1937, p. 231).

Turing was aware of what he was accomplishing. He provided, as Church put it later, an intuitive notion of computation that becomes immediately clear to the reader—as opposed to, say, the one given by his own lambda-calculus. Turing exhausted the realm of possibilities of what a computation could do—at least in this intuitive way—by this machine: "It is my contention that these operations include all those which are used in the computation of a number".[15] Further, he also gave an algorithm (the set of instructions) the absolute power over its effected procedure: it would completely determine the behavior of the chosen parsed area of reality...

> We may compare a man in the process of computing a real number to a machine which is only capable of a finite number of conditions q1, q2,..., qR which will be called "m-configurations"...The possible behaviour of the machine at any moment is determined by the m-configuration qn and the scanned symbol S(r). This pair qn, S(r) will be called the "configuration": thus the configuration determines the possible behaviour of the machine.[16]

To recall, the main challenge was to find out whether there was an effective procedure by which one could demonstrate, in finite time, whether a given mathematical proposition is provable from a finite set of axioms. It soon became evident that the notion of an "effective procedure" should be defined—or at least clarified. As mentioned above, Alonzo Church identified an effective procedure with an "algorithm", thus making inroads toward its definition. And we have seen how Turing's insights into the nature of an algorithm, within the broader aim of answering Hilbert's challenge, coincidentally encountered the parallel development of Church's work. As also mentioned, Church beat Turing to the punch in publishing similar findings

---

[15] Ibid., p. 232

[16] Ibid., p. 231. The reader may have noticed the introduction of the notion of "man" for the first time, alluding to ideas of intelligence and behavior. The issue of whether or not a machine can be intelligent (*contra* Turing at this point) was to emerge still some time in the future. Let us recall that computers as we know them did not yet exist at this moment. This issue will, however, come back below— Sections "Alan Turing's Strange Reversal Regarding the Question of Artificial Intelligence" & "Cybernetics as a Possible Missing Link Regarding Turing's Change of Heart" of this chapter.

some months before the latter was given the *placet* by his supervisor, Max Newman, for submitting his own piece in 1936—and published the next year. However, by recommendation of his own supervisor, Turing was by the end of the same year at Princeton University, pursuing doctoral studies under Alonzo Church himself.

Two years were enough for Turing to obtain his Ph.D., with a dissertation entitled "Systems of Logic Based on Ordinals".[17] This monograph, often historically overlooked, built upon the idea of the noncomputability of certain realms of mathematics. Within this framework, the thesis left important cues regarding Turing's own later theoretical evolution. More relevantly to the issue that concerns the present work, Turing's dissertation contains seminal insights into what he would consider to be a machine. In this thesis, in an excerpt that could be taken as joint statement regarding the equivalence between the definitions of computability reached by Gödel, Church and himself, Turing wrote:

> A function is said to be "effectively calculable" if its values can be found by some purely mechanical process. Although it is fairly easy to get an intuitive grasp of this idea, it is nevertheless desirable to have some more definite, mathematically expressible definition. Such a definition was first given by Gödel at Princeton in 1934...These functions were described as "general recursive" by Gödel. We shall not be much concerned here with this particular definition. Another definition of effective calculability has been given by Church..., who identifies it with lambda-definability. The author has recently suggested a definition corresponding more closely to the intuitive idea...[18]

Turing then puts emphasis on the identification of effective calculability with a mechanical process, equation that permits the other two definitions (Gödel's and Church's) to also emerge. Here Turing came close to identifying what a machine might stand for,[19] doing it in a

---

[17] Turing 1939.

[18] Turing 1939, p. 166.

[19] Close, but not quite there. Hodges acknowledges that the "central thrust of Turing's thought was that the action of any machine would indeed be captured by classical computation", and *that* is the closest that Turing gets to defining a machine (Hodges 2008, p. 88). Hodges, however, reduces Turing's probable meaning of "machine" to "Turing machine" (Hodges 2008, p. 77).

reverse manner: if a process is said to be "purely mechanical" then one is to expect that there is a machine carrying that process. "Purely mechanistic" implies a machine doing the job.[20] Turing's thesis states that

> "a function is effectively calculable if its values can be found by some purely mechanical process". We may take this statement literally, understanding by a purely mechanical process one which could be carried out by a machine.[21]

This machine, bounded by its own nature, exhaustively generates something germane to its own being: A mechanical process. In fact, the analytical relation between a concept and an idea already contained in the concept (e.g., a triangle has three sides) is not foreign to this relation. A mechanical procedure is already contained in the idea of a machine. The machine, emanating what is to be understood as mechanical *per se*, completely determines what the mechanical procedure will be.

The suggestion of having a model of machines—a paradigm of what can exhaustively perform a mechanical process—where its physicality is irrelevant, did not go unnoticed. Indeed, the "purely mechanical process"

---

[20] When Turing talks of a machine, right from his 1937 paper onwards, he is implicitly talking about an "automatic" machine—as opposed to a "choice" machine (a machine whose instructions and outcome leave room for human tweaking) or an "oracle" machine (a fabled machine which processes uncomputable realms). Each would be named a-machine, c-machine and o-machine, respectively. He stated that when he will refer to a "machine" he will always be referring to an a-machine. So he conventionally dropped the apposition "a-" and begun to call them simply "machines" (Turing 1937, p. 232).

[21] The quote continues, making explicit the reliance on the articulation of an effective calculability (computability) by means of making it amenable to being instantiated as a mechanical procedure:

> It is possible to give a mathematical description, in a certain normal form, of the structures of these machines. The development of these ideas leads to the author's definition of a computable function, and to an identification of computability with effective calculability. It is not difficult, though somewhat laborious, to prove that these three definitions are equivalent (Turing 1939, p. 166).

here is instantiated by an algorithm[22] (a set of commands)—"a machine". What makes a machine ultimately a machine, even if it is not yet defined, is just tangentially related to its possible physicality. The latter is not referred to as a *sine qua non* condition pertaining to its nature. Instead, what makes a machine to be itself is its inherent capability of producing a determinate and effective procedure, a "purely mechanical" outcome.

There had certainly been in history several metaphysical appreciations of mechanical entities. The Scientific Revolution emphasized mechanisms, beyond experimentation, as a sure-proof way for building up grounded theoretical edifices.[23] Rene Descartes embodied this lemma with an intensity that is often overshadowed by his legacy of "substance dualism". A less often noticed side of his philosophy was his faithful reliance on a literally mechanical view of organisms. Historically recalling, Descartes divided reality in two qualitatively distinct substances: *res extensa* and *res cogitans*. The first one referred to things that had materiality (physical extension), for example, rocks, plants, animals, human bodies. The second one referred to a substance, equally existent and real as the first one, but which lacked physicality: minds, angels, and God.[24] The theological benefits of this separation have been often pointed out as a purported move to

---

[22] "Turing's work gives an analysis of the concept of 'mechanical procedure' (alias 'algorithm' or 'computation procedure' or 'finite combinatorial procedure')." (Gödel 1934, p. 72)

[23] Paradigmatic in this reliance upon a mechanical element of knowledge is Francis Bacon's praising of the "mechanical arts," collapsing the Greek legacy of a distinction between practical and theoretical knowledge. Wisdom shall now be merged with mechanical mastery, culminating in Bacon's *experimentum crucis*. Indeed this distinction (particularly the Helenic frowning upon τέχνη) is what allegedly crippled the development of science—as Bacon understood it—among the Greeks...

> Whereas in the mechanical arts, which are founded on nature and the light of experience, we see the contrary happen, for these (as long as they are popular) are continually thriving and growing, as having in them a breath of life, at first rude, then convenient, afterwards adorned, and at all times advancing (Bacon 1620a, § 74).

[24] "I deny that true extension, as it is usually understood by everyone, is found either in God or in angels or in our mind, or, in short, in any substance that is not a body" (Descartes 1649, p. 293).

"protect the soul", given the advances of the new sciences. But another alleged advantage gained from this was the possibility of denying outright any animal participation of soul properties, avoiding the ontological hassle of distributing watered down versions of the *logos* throughout the animal kingdom (from primates all the way down to insects and beyond). In this way the special place of man in the universe, as image and likeness of God, gets removed from an otherwise dangerous situation of being superior only as a matter of degree, and not in kind. Another claimed benefit was the continuation of the tradition of guiltlessly practicing vivisection—following the Greek physician Galen and Leonardo Da Vinci, who reported important insights into physiology as an outcome.[25]

The soul is far removed from this picture in its quality of *res cogitans*, and thus, safe. However, both animals and human bodies pertain to the realm of an organic *res extensa*, and thus, they are totally subjected to the deterministic laws of mechanics, which could now be pursued. On his *Treatise of Man*, Descartes refers to the intricacies of the human body, with all its vegetative and non-vegetative functions, in the following terms:

> I desire, I say, that you should consider that these functions in the machine naturally proceed from the mere arrangement of its organs, neither more nor less than do the movements of a clock, or other automaton, from that of

---

[25] The fact that these were advances coming from Protestant realms was probably also part of his motivation. There is a received view that Protestant colleges were putting emphasis on the development and teaching of the sciences and "liberal arts" and trades, whereas the more traditional Catholic universities were still emphasizing the teaching of ancient languages, philosophy, theology, and the fine arts. The Church purportedly preferred a unified Europe under its tutelage, uniformly communicating via Latin. Descartes' academic formation however came from a school set up in the context of the Counter-Reformation. The Jesuits instilled in him a deep admiration for mathematics and sciences, while studying at the Collège Royal Henry Le Grand at La Flèche. Descartes' Catholicism, although not particularly orthodox, was likely profound. The abdication of Queen Christina of Sweden, his pupil, in order to become a Catholic, is often brought up as an instance of Descartes' faith all the way to his death (Swedish law, as in today's UK, prohibited a Catholic to become a monarch). For a recent study on Descartes' intellectual life, including the context of his education and religious beliefs, see Gaukroger 1995. For an account of the educational context of Europe after the Reformation, see Eisenstein 2012, ch. 7.

its weights and its wheels; so that, so far as these are concerned, it is not necessary to conceive any other vegetative or sensitive soul, nor any other principle of motion or of life, than the blood and the spirits agitated by the fire which burns continually in the heart, and which is no wise essentially different from all the fires which exist in inanimate bodies.[26]

As one can readily notice, what is relentlessly mechanized is still the realm of material entities—*res extensa*. The demystification that is taking place is circumscribed to stripping the organic area of *res extensa* from any purported vitalism inherent to the phenomenon of "life". Clocks are as much automata as animals and human bodies are. But the mind, inhabiting the realm *of res cogitans*, is still perfectly safe—anticipating Kant's anxiety regarding human freedom for almost 150 years.[27] In fact, the mechanization of something other than *res extensa* would probably not have made much sense for Descartes.[28]

[26] Descartes enumerates in the following paragraph what functions he is referring to:

"All the functions which I have attributed to this machine (the body), as the digestion of food, the pulsation of the heart and of the arteries; the nutrition and the growth of the limbs; respiration, wakefulness, and sleep; the reception of light, sounds, odours, flavours, heat, and such like qualities, in the organs of the external senses; the impression of the ideas of these in the organ of common sensation and in the imagination;… the retention or the impression of these ideas on the memory; the internal movements of the appetites and the passions; and lastly the external movements of all the limbs, which follow so aptly, as well the action of the objects which are presented to the senses, as the impressions which meet in the memory, that they imitate as nearly as possible those of a real man." (Descartes 1662, § 202)

[27] Descartes originally wrote his *Treatise on Man* in 1633, but it was posthumously published in 1662. Kant wrote his first two *Critiques* in 1781 and 1787, where he laid out his grand project of saving freedom by restricting the realm of a rightfully mechanistic Modern science to the *phenomenal*—sparing the *noumenal* realm for our exercise of free will.

[28] Somewhat anachronistically, Michael Wheeler speculates that Descartes could have expanded his notion of the mechanical to the non-physical (*res cogitans*), but that his view of a machine was, due to his seventeenth-century context, somewhat crude (Wheeler 2008). The mechanization of the whole of man (not only of his body), a related but different issue—where Julien Offray De La Mettrie (1709–1751) would be relevant—lay ahead in the near future.

In contrast, with Alan Turing we have the notion of a non-physical entity as a certain type of machine—a non-material one. This never happened before. It had not happened because the notion of a machine had not yet been developed to a point at which it would give up its physicality as a non-essential feature. Although the horizon of this insight might have been already simmering for a while, it was not made explicit until the above addressed evolution of logic and mathematics took place. That development was the articulation of the notion of an algorithm—explicitation deemed necessary for fully clarifying the basic notion of an effective computation. As previously mentioned, this definition was in turn necessary for addressing Hilbert's dramatic question regarding the possibility of producing an algorithm that would solve any mathematical problem within a finite time and within a finite set of axioms. Both Turing and Church attempted to define such a notion (of an algorithm), and the isomorphic correspondence to a familiar object in the physical world was soon evinced. That object, to rehash, was a machine. More specifically, a physical machine is but one instantiation of what a machine is *per se*, since physicality is *not* a *sine qua non* feature of a machine *qua* machine. With this new understanding of a machine, devoid of a heretofore deemed necessary physicality, previous realms hitherto beyond the scope of a rigorously mechanistic explanation were suddenly amenable to being encompassed. Such as life and mind.

Something remarkable—over and beyond the response to an admittedly important mathematical problem—was thus accomplished in Turing's "On Computable Numbers". Although no definition of machine was given in the article (or in any other writing by Turing for that matter),[29] the seeds of an insight that would prove powerful for subsequent scientific, technological, and philosophical endeavors in decades to come were already there. Turing stopped short of defining a machine, but the rigorous explicitation of what stands for a "mechanical procedure" demanded from him the abstract construction of a machine that would do precisely that. This abstract machine, capable of exhaustively (and effectively) performing any possible computation (as construed by Turing), became a paradigmatic machine. An ideal machine, model of all machines. Again, nothing had so far put forward the notion of

---

[29] Turing might have been close to disclose a full-fledge theory of machines; but he never did. See the beginning of this section.

a non-physical machine. Hence this observation was rather quickly noticed by some scientists, mathematicians, psychologists, and thinkers alike, expanding their foreseeable theoretical exercises and scientific playgrounds. Here we come to a realization of something that, although probably still unaware of its full impact, might have changed the course of recent history. This is one of the pillars upon which a scientific movement that lasted a decade (1942–1952) was built. Turing's insight into the nature of an algorithm, an essential feature of which being that it maintains an isomorphic relation with an entity in the physical world—namely, a machine—can be regarded as the theoretical foundational bedrock upon which this was devised as a general scientific theory. The name of this scientific outlook went under the name of *cybernetics*.

ALAN TURING'S STRANGE REVERSAL REGARDING THE QUESTION OF ARTIFICIAL INTELLIGENCE

We have seen how the need for a clarification of the notion of an algorithm was a pre-condition for defining the idea of a computation, with the goal of addressing Hilbert's problem. We have also seen how this task opened up insights into the nature of a machine, first captured, although not fully worked out, by Alan Turing in his "On Computable Numbers". There is another realm, however, in which cybernetics and Turing might have been deeply theoretically intertwined as well. The question of Turing's reversal regarding the possibility of a machine possessing intelligence has been tangentially addressed, but the puzzle regarding this dramatic turn seems to remain. This puzzle would involve finding out what went through Turing's mind to make him change his view on the philosophically important issue of Artificial Intelligence (AI).

The development of cybernetics might shed some light on what went through Turing's mind for the span of a decade, probably pointing to the reason why he evolved into an early advocate of AI, epitomized in his 1950 article "Computing Machinery and Intelligence".[30] At some point between 1936 and 1946, Turing's views regarding what a machine could do underwent a strange reversal. In fact, the contrast between what Turing

---

[30] Turing 1950.

would say regarding the possibility of machine intelligence[31] in 1936, and then almost 15 years later is startling. Let us return to the interesting remark of his 1937 article[32]:

> We may compare a man in the process of computing a real number to a machine which is only capable of a finite number of conditions q1, q2, ..., qR which will be called "m-configurations"... The possible behaviour of the machine at any moment is determined by the m-configuration qn and the scanned symbol S(r). This pair qn, S(r) will be called the "configuration": thus the configuration determines the possible behaviour of the machine.[33]

The reader will surely note the quiet comparison between a man and a machine. This analogy has triggered some confusion in the past. Let us be reminded that the existence of computers, as we know them now, still lay somewhat far in the future, at least one decade later. At the time, however, the task of computing numbers already existed—in fact, it always did.[34] The task of crunching numbers with a pen and paper was part of the idiosyncratic picture of available jobs at the time (indeed usually held by women). This job, infamous for its tediousness, required the blind following of a rule, absolutely and without deviation. In the same vein, Ludwig

---

[31] "Machine intelligence", in this context, refers to straightforward human intelligence inhabiting an artificial milieu, i.e. a machine—as we will see below. The nature and evolution of the notion of intelligence (which allowed its usage in, for example, contemporary electronic engineering) is beyond the scope of this work.

[32] Quoted in the previous section.

[33] Turing 1937, p. 231.

[34] There is no universally agreed definition of computation. The intuitive depiction of it by Turing gained traction in what regards the future instantiations of computing machines. But certainly there are other ways to set a computational process. That raises the question whether the existence of "mind" is a necessary prerequisite for a computation to take place at all—someone has to perform the computation, for example, make up the set of instructions and run it. After the arrival of the "genetic revolution" some would contend that computation occurs in nature itself—after all, it is a "mindless" procedure. Some would suspect that it may even subsume the entirety of physics into a realm too immense for us to grasp at this time. For an attempt to lay out a panoramic vista of theory of computation in our days—friendly to the view just mentioned—see Smith 2002.

Wittgenstein has a relevant remark that reeks of mysticism only if taken without the above context: "Turing's 'Machines'. These machines are *humans* who calculate".[35] The *machinic*[36] character of jobs of such nature (e.g. the endless and monotonous repetition of a small task or movement), where "mechanical" and "mindless" were referred to as synonymous, was a common theme in the popular culture of the time.[37] In fact, the media began to refer to computing machines as "computors" in order to differentiate them from "computers", the latter being humans who had the hard and boring job of building up number tables. As machines gradually (and literally) took over the job, they were eventually referred to as "computers".[38] Turing's 1937 paper, therefore, should not be regarded yet as a precursor to what later came to be known as "Artificial Intelligence". Not only were there no computers (in the modern sense) at the time, with which humans could be compared, but also the very notion of "mechanical" carried an epithet of "mindlessness" inherently attached to it. This pejorative adjective ascribed to a machine was accepted and defended by Turing himself—which would make the notion of "machine intelligence" an oxymoron, and by extension "artificial intelligence" senseless.

Further, the paper had as its original aim answering Hilbert's Decision problem, and since this answer was given in the negative, the paper showed that some mathematical propositions are in fact not computable—hence the answer *must* be in the negative. Some incomputable realms of mathematics in essence cannot be treated by a machine. This limitation as to what a mechanical procedure can do puts a machine further away from a possible comparison with the human mind, canonically agreed to be capable, by Turing's own standards, of dealing with non-mechanical realms, as it is the case with "intuition"—indeed a familiar occurrence

---

[35] Wittgenstein 1980, §1096.

[36] The term—at least as applied to this context—is John Johnston's (Johnston 2008).

[37] Tellingly, Charles Chaplin's iconic film "Modern Times" was released in 1936. A vestigial sentiment of repulsion triggered by the dehumanizing social effects of the Industrial Revolution might have been at play.

[38] Thus Wittgenstein's remark could be transduced as "Computers are computors who calculate", or simply, computers do the mechanically mindless job of a human, faster.

reported among mathematicians like him while engaged in problem solving. Turing's change of mind, giving a machine a qualitatively new kind of cognitive powers, still lay ahead in the future.

After Turing had defended his dissertation, he was offered a position at Princeton, but he refused it in order to go back to Britain. The shadow of war was already looming and it is quite possible that he was moved by patriotic considerations. From 1939 (year in which his dissertation was published) to 1945, Turing was submerged in war-related tasks—the secrecy of which partly explaining a publishing hiatus. By 1945 the design for an early electronic computer, the Automatic Computing Engine (ACE) was finished: a task that demonstrated, beyond his mathematical brilliance, also his engineering skills.

Tinkering with machines was not an altogether new experience for Turing—it is reported that he had an uncanny attraction toward machinery since his childhood (for example, with the nature of a typewriter). During the span of this decade (1936–1946), something occurred in Turing's mind regarding the status of machines, the computable and intelligence. In 1946, in a report on the ACE, he suddenly (and mellifluously) introduced for the first time the idea of "mechanical intelligence"—a notion regarded at the time as radically senseless. But he did it in such a tangential, almost cryptic way, that it seemingly did not raise many eyebrows (which could be partly understood given the ambience of secrecy of the era)...

> We stated...that the machine should be treated as entirely without intelligence. There are indications however that it is possible to make the machine display intelligence at the risk of its making occasional serious mistakes.[39]

Advancing an insightful premise, it was suggested that making mistakes could be a sign of intelligence. Moreover, one could certainly program a computer in a way that it could "make the machine display intelligence".[40] Turing did not elaborate much on the bone thrown there. Why would mistake-making be a reproducible sign of intelligence? In 1947, in a presentation for the London Mathematical Society, Turing expanded on the "bizarre" idea of "machine intelligence", again touching the issue of what mistake-committing might entail for a machine. Connecting with

[39] Turing 1946, p. 16
[40] Ibid.

previous work done in the foundations of mathematics, he described what the outcome is for both a machine and a mathematician dealing with paradoxes encountered in logic and arithmetic. By the very nature of mathematics, according to the recent discoveries by Gödel, Church and himself, there cannot be an algorithm that would find out whether *any* mathematical proposition is true or false—indirectly answering Hilbert's problem. That entails that a machine will eventually stop in its attempt to prove a certain equation, in a particular system—one in which there will be at least one true proposition that is not provable...

> It has for instance been shown that with certain logical systems there can be no machine which will distinguish provable formulae of the system from unprovable, i.e. that there is no test that the machine can apply which will divide propositions with certainty into these two classes. Thus if a machine is made for this purpose it must in some cases fail to give an answer. On the other hand if a mathematician is confronted with such a problem he would search around and find new methods of proof, so that he ought eventually to be able to reach a decision about any given formula.[41]

Given the undecidability of mathematics (even given the non-facile nature of mathematics, period), humans commit errors in looking for solutions to complex problems. But we still call them "intelligent". The "intelligent" mathematician will look for other ways, attempt other strategies, in order to continue the task. However, when it comes down to machines, the pre-conceived notion exists that "if a machine is expected to be infallible, it cannot also be intelligent".[42] Turing denounces that this is "unfair play" to the machine.

> I would say that fair play must be given to the machine. Instead of it sometimes giving no answer we could arrange that it gives occasional wrong answers. But the human mathematician would likewise make

---

[41] Turing 1947, p. 497.

[42] Ibid. We will see that the emanation of something as exact as *mathematics*, from something as messy as the *mind*, disturbed other thinkers. Superimposing the former—amenable of instantiation by a machine—upon the latter, in order to explain mental behavior, will have far-reaching implications. See the case of Warren McCulloch in Chapter 3, Section "The Decline" and Chapter 5, Section "Machines can be Immaterial" of this work.

blunders when trying out new techniques. It is easy for us to regard these blunders as not counting and give him another chance, but the machine would probably be allowed no mercy.[43]

If, somewhat ironically, the mistake-committing human is regarded as intelligent, whereas the infallible machine is not, then let us make the machine become fallible. A machine that commits mistakes will then display a feature so far reserved only for humans—hence Turing 1946 remark that "it is possible to make the machine display intelligence at the risk of its making occasional serious mistakes".[44] This allowance for mechanical mistakes opens up another realm of theorizing about the possible intelligence of a mechanical entity. After technology advances a fair bit (Turing wonders about a future with machines bearing a substantially expanded "storage capacity") nothing would in principle preclude the machine from changing its own set of instructions. Therefore, the machine would be allowed to commit errors so that it could be able to reorganize itself, even if for this "the machine must be allowed to have contact with human beings in order that it may adapt itself to their standards".[45] This suggestion, namely, that of a machine self-changing its internal parameters, carves up still another area of comparison between a machine and a human being—this time with a child, in the context of the infant's learning process:

Let us suppose we have set up a machine with certain initial instruction tables, so constructed that these tables might on occasion, if good reason arose, modify those tables. One can imagine that after the machine had been operating for some time, the instructions would have altered out of all recognition, but nevertheless still be such that one would have to admit that the machine was still doing very worthwhile calculations. Possibly it might still be getting results of the type desired when the machine was first set up, but in a much more efficient manner. In such a case one would have to admit that the progress of the machine had not been foreseen when its original instructions were put in. It would be like a pupil who had learnt

[43] Turing 1947, p. 497
[44] Turing 1946, p. 16.
[45] Turing 1947, p. 497

much from his master, but had added much more by his own work. When this happens I feel that one is obliged to regard the machine as showing intelligence.[46]

The picture of a creature surprising its creator with its behavior has been a common Golem-type theme among humanists worried about the possibility of "playing God".[47] Self-modifying algorithms could in principle, eventually, remove from the machine the stigma of being a mere instruction-following brute. However, this would be just a first step toward the mechanical appropriation of the notion of "intelligence". Turing was aware of this, and he went on to address the notion of "learning", revealing quite evident behavioristic cues.[48] Thus in his 1948 report "Intelligent Machinery", the iconoclastic comparison of a "learning machine" with a "learning child" was further elaborated.

Firmly entrenched into the grand-idea of a Mechanistic Continuum[49] underpinning much of Modern science and philosophy, Turing was well aware of the foundational experiments in behaviorism performed by the Russian physiologist Ivan Pavlov (1849–1836) and his followers. The phenomenon of "learning", when broken down into pieces, reveals a "mechanism" (as anything amenable of scientific treatment, for that matter): a building and reinforcement of neural pathways, conditioned by external stimuli. This mechanism can be ascribed across the natural spectrum to the lower forms of life, as well as to the higher vertebrates.[50] Any spiritual, irreducible, epiphenomenal understanding of "learning" that recoils against this "empirical" explanation of what actually occurs behind the phenomenon (namely, its "mechanics"), would merely be the death rattle of a metaphysically ridden Medieval understanding of man and nature.

In the same vein, for Turing a child's cortex "is an unorganized machine, which can be organized by suitable interfering training".[51] The

---

[46] Ibid., p. 496
[47] This is a main theme in Wiener 1964.
[48] See Shanker 1995.
[49] Ibid., p. 53.
[50] Ibid.
[51] Turing 1948, p. 424.

training of the child, the "teaching", is not different in principle than those occurring in Pavlov's dogs, laboratory worms, and "evolving" machines. Learning pathways get created, and then reinforced, when associations between a certain state and a stimulus of pleasure (or pain) are repeatedly allowed to occur. The "unorganized machine" with the capacity of altering its own instructions (with a storage capacity that would provide enough "memory"), would be equally suitable for showing off real learning, after a period of productive interaction with a (probably human) master who has delivered the appropriate punishments and rewards. Three years later, in a talk similarly entitled as the report, albeit with a bolder twist—"Intelligent Machinery, A Heretical Theory"—Turing was more specific regarding the behavioral training of this "child machine":

> I suggest that there should be two keys which can be manipulated by the schoolmaster, and which can represent the ideas of pleasure and pain. At later stages in education the machine would recognize certain other conditions as desirable owing to their having been constantly associated in the past with pleasure, and likewise certain others as undesirable.[52]

The next year Turing began to write the seminal AI paper "Computing Machinery and Intelligence" (to appear in 1950 in *Mind*, the only philosophy journal where he published throughout his career—a testament for how his ideas got wide circulation, attracting the attention of the philosophical community). This paper has been one of the most quoted philosophy papers from the second part of the twentieth century on, so I will not dwell on it here. Suffice it to say that to a great extent this paper put together in an organic manner the ideas put forward by Turing since 1936—but especially since after the war, where his commitment to a total mechanization of intelligence was clear. For instance, the famed "Turing Test", subject of much philosophical debate up to these days, can already be foreseen in the 1946 paper (the ACE report). There Turing writes about paying the price for "seeing" intelligence in the machine by allowing it to have errors, suggesting that "it is possible to make the machine display intelligence at the risk of its making occasional serious mistakes".[53]

---

[52] Turing 1996, p. 259. Although the talk was given in 1951, it was published posthumously in 1996.

[53] Turing 1946, p. 16

We are here in front of a full 360 degree turn in what concerns the possibility of a machine being intelligent—from its denial, based upon the mechanical incapacity of dealing with intuition, to its defense, based upon the possibility of a machine displaying human features, such as mistake-committing and learning. Since arguably much of contemporary cognitive science pivots upon the notion of "artificial intelligence" as articulated in this 1950 paper, the question still remains regarding Turing's change of mind. What made him go from the position where mechanical procedures merely were those underlying the accomplishment of tasks by instruction-following brutes, to the position where these very mechanical processes could expose true intelligence in those bearing them? One can rephrase the question factoring in the historical framework. What went on with Turing behind his gradual dismissal of the previously vigorously defended[54] "oracle machine"—the fabled machine that could compute the uncomputable? This first position was changed later on, where not only the discrete areas of thought dealing with mathematical problems but also the whole of human intelligence, was to be understood as being essentially computational in nature. Oracle machines were no longer necessary to account for those areas that cannot be subsumed to a machine, since everything was mechanical in one form or another, for example, even via subconscious computations. Now, intelligence could be one day artificially constructed.

Cybernetics as a Possible Missing Link regarding Turing's Change of Heart

There are those who find reasons of a personal character concerning Turing's change of heart. Alan Turing had an early love in his teenage years who died prematurely of tuberculosis. This death haunted him for the rest of his life. It has been speculated that a subconscious wish of bringing back the beloved one (or making a beloved one never die) was behind the relentless desire for making the whole of intelligence mechan-izable—and thus, constructible and retrievable. Turing handwrote in 1932 a short essay on the "Nature of the Spirit" where he wonders after

---

[54] In his doctoral dissertation (Turing 1939).

his friend's death whether a detached spirit can re-inhabit another adequate vessel:

> ...as regards the actual connection between spirit and body I consider that the body by reason of being a living body can "attract" and hold on to a "spirit", whilst the body is alive and awake the two are firmly connected. When the body is asleep I cannot guess what happens but when the body dies the "mechanism" of the body, holding the spirit is gone and the spirit finds a new body sooner or later perhaps immediately.[55]

There seems to be the hope that a mind could exist without its biological brain, probably amenable to be re-instantiated once a proper holding mechanism is achieved.[56] This would not seem too farfetched from Turing's mechanistic views if we understand both realms, the material and immaterial, as being equally mechanical—and hence, with a nature determined not by physicality, but by mechanicity. Turing's later strong correlation between the notion of an algorithm and that of a machine might have been a cornerstone toward a further enrichment of his notion of a machine—culminating in the oracle machine (below) where determinate-behavior as an inherent feature of a machine is superseded.[57]

There are others, more lenient toward a "Science and Technology Studies" view, that give a prime role to Turing's confrontation with engineering feats during the war, which allegedly affected his subsequent views. Turing saw first-hand the power of machines, designed and constructed by him to break the German code of the sinister U-Boats.[58] For Winston Churchill, this accomplishment greatly

---

[55] Hodges 1983, p. 63.

[56] Jean Lassègue proposes that the later Turing Test had this psychoanalytical motivation (Lassègue 1996).

[57] Andrew Hodges disagrees (Hodges 2000). "On Computable Numbers" marks for Hodges the end of any dualistic remnant in Turing's mind, committing the mathematician for good to a deterministic and materialistic stance. It is relevant to remark that the case could be made that Turing's 1) later oracle machine; and 2) insights into the machinal as anchored in mechanicity and not materiality (as depicted in the 1937 article and his 1939 dissertation) could attenuate Hodges' objections.

[58] Hodges shares this view—although he does not give much of an articulation of the alleged link between Turing being impressed by powerful machines and the possibility of machine intelligence (Hodges 2000).

contributed toward Great Britain's victory,[59] acknowledging Turing's essential contribution toward the Empire's victory[60]—a victory that might not have happened if the calculations would have been performed solely by humans. Thus, Turing witnessed the rapid improvement in the construction of these all-important machines, going from entirely re-wiring the machine to vacuum tubes and from there to relays—one design material making the machine faster and smaller than the next one. What could the future hold in 50 years from then? That is the question that Turing poses toward the end of his 1950 paper, and he is of the idea that intelligence will be mechanical.[61] And this would not be our intelligence, but the machine's: it would be *artificial intelligence*.

Still others, more philosophically oriented, in a view not altogether unrelated to the previous one, see in Turing's change a gradual but firm realization of his own place in the grand-scheme of the above mentioned Mechanistic Continuum. If he was to bring his own insights—already firmly set in such a tradition—to their rightful completion, then thinking machines would necessarily follow. Learning, or the acquisition of knowledge, as indicated above, can be broken down into pieces that are so small that they can be found neatly interacting with the overall machinery of survival, even in the simplest organisms. Relevant connections for survival are forged and reinforced via trial and error, the environment producing stimuli in the form of reward or punishment. Certainly the learning process of a worm and of a classical composer can be regarded as vastly different, but this difference would not be one of kind, but of degree. The latter is greatly more *complex*, but the principle of productive change upon which both rest is largely the

---

[59] Turing and his fellow co-workers wrote a conjoint letter directly to Churchill, where they vented their frustrations due to the slowly bureaucratic pace at which their logistic demands were being met. Upon receiving the letter, Churchill wrote a terse and short note to his Chief of Staff: "ACTION THIS DAY. Make sure they have all they want on extreme priority and report to me that this had been done." (Copeland 2004, pp. 336–340).

[60] "According to Winston Churchill, Turing made the single biggest contribution to Allied victory in the war against Nazi Germany." (Spencer 2009).

[61] Turing 1950, pp. 455–456.

same.[62] This realization might have further laid out in Turing the foundation for the hope of an artificial intelligence to emerge.

All these interesting reasons notwithstanding, there nevertheless is an avenue of possible explanation for Turing's reversal that has been less explored—but which popped out as a recognizable clarifying scheme during this writer's research. This explanatory path goes through the historical cross-roads that one could identify within the decade mentioned above (1936–1946). Between Turing's insights characterizing a machine as being inherently incapable of intelligence, and the now ubiquitous AI-friendly classical computationalism stemming from Turing's later position, there is a missing link. To explore this suggestion, we have to go back to the time frame occurring between Turing's 1937 paper and his subsequent thesis dissertation under Alonzo Church. Laying out this line of explanation, which emerged in the midst of my own investigation into the nature of a machine, will hopefully open up the context for introducing a theory in the history of science that revolved around the notion of a machine *tout court*—cybernetics.

Turing's paper "On Computable Numbers", as previously mentioned, contains the seed of a radical view of a machine, which seized the attention of a handful of scientists—who were soon to engage in the cybernetic enterprise. Turing, to recall, further elaborated the notion of a machine for his doctoral dissertation under Alonzo Church. However, instead of staying in America, he chose to return to England, likely motivated to help his country *vis à vis* the looming war on Europe.[63] Once back in England,

---

[62] "Herein lies the thrust of the behaviourist version of the continuum *picture*... The crucial point is the idea that *learning* consists in the formation of stimulus-response connections which themselves require no intelligence...The 'higher' forms of learning...are distinguished from these lower forms by the complexity of the stimulus-response connections forged in the organism's brain. But the mechanical nature of the atomic associations which form the basis for all levels of learning remain identical. This provides the rationale for describing what had hitherto been regarded as disparate phenomena...as constituting a *learning continuum* ranging from simple negative adaptation, habituation, accommodation, and tropisms, through animal and infant learning, to the highest reaches of education and scholarship" (Shanker 1995, p. 63).

[63] One cannot help wondering about the counterfactual of Turing having stayed in America, and in all likelihood having joined the "cybernetic group".

Turing was a regular attendee at the British version of the Macy Conferences—the much smaller and informal Ratio Club.[64] Both gatherings, Macy's and Ratio's, exchanged some speakers—mostly coming from the USA to the UK. That is how Wiener and McCulloch showed up in England on occasion. Turing, later on, while engaged in the war effort, would go to the USA to share knowledge regarding the construction of the novel computing machines. Even if he did not ever attend a Macy meeting, he was in constant contact with the cyberneticians—especially with von Neumann, due to the common task of building the first computers—who would act as hosts during his stay in America.

In other words, although Turing left aside further developing the notion of a machine *per se*—and engaged instead in implementational endeavors emanating from it, such as breaking the German-encrypted communications by means of the urgent construction of a computing machine—he was still in proxy contact with the cybernetic enterprise, running strong in the USA. Ten years later, when Turing wrote the 1946 report where he muses about the possibility of a thinking machine, cybernetics was reaching its own pinnacle. The cybernetic view of a machine was much further elaborated, and the notion of a de-physicalized mechanical entity already had full and rightful existence.[65] The mind was already a machine. So when someone like Turing would ask, in his own terms, the heretical question "can a machine think?", the answer, coming from cybernetics, could only had been in the positive: they obviously can—they already do! The existence of our mind is there for proof. Ours *is* a mechanical intelligence.

This slow but firm parallel development, which ended in a reversal of the question, might have dynamically exercised influence upon Turing himself, in order for him to state his famous rhetorical question.[66] It is

---

[64] See Chapter 3, Section "The Ratio Club".

[65] In all fairness, due in part to pre-cybernetic British theoretical influence. See the section on Kenneth Craik in Chapter 5, Section "Machines are Embodied Theories" of this work.

[66] Jean Pierre Dupuy sees a connection between AI and cybernetics in this respect, but he does not place it specifically on Turing (Dupuy 2000, Chapter 5). Indeed, Piccinini criticizes Dupuy's book for failing to fully flesh out the alleged direct (albeit fairly unrecognized) cybernetic ancestry of cognitive science as a whole (AI including)—which is seemingly promised at the beginning of the book (Piccinini 2002).

not farfetched to suppose, then, that cybernetics' own nuanced development of the nature of a machine (encompassing non-material entities through and through) might have played a non-minor role in Turing's evolution regarding the capabilities of mechanical entities—indeed particularly regarding thinking. Just as the computer was an implementational spin-off from the attempt to effectively logicize the mind,[67] cybernetics was just as much a spin-off from the attempt to understand the nature of computability—articulating the nature of a machine. But while Turing seemingly wanted to give machines anthropomorphic features, cybernetics wanted to make of man a machine.

Artificial Intelligence left aside further investigation into the nature of a machine—and marched forward instead with the anthropomorphization of a machine (they would want a machine to show a human feature: thinking). A key for understanding the theoretical launch of AI would, thus, be Turing's late change of mind in terms of what machines could do. First, they could not ever show capacities reserved only for humans—such as intuition. Later, in principle, they could do just that. The subsequent development in Turing's thoughts not only could help understanding why cybernetics ended up being a somewhat "fringe" occurrence in science history, but could also help clarifying the reason behind the overemphasis on classical computationalism in today's cognitive science.[68]

A novel science, cybernetics, might have deeply influenced the mathematician Alan Turing toward a profound change in perspective regarding a

---

[67] John von Neumann based his famed "First Draft" upon McCulloch and Pitts'1943 article for designing the architecture that underlies most computers today, known now as the von Neumann architecture. This claim is not free from controversy. More on this in Chapter 7.

[68] Although not fully articulated, Dupuy correctly recalls some historical pointers that lead toward the orphan status of cognitive science as a whole—reluctant to acknowledge its cybernetic origins (Dupuy 2000, ch. 2). In all fairness, exposing this historical (and philosophical) ancestry might merit a doctoral dissertation on its own. This issue is not unrelated to the historical episode depicting an incipient research in Artificial Intelligence aggressively competing for funding at the beginning of the Cold War (See Chapter 3, Section "Traditional Explanations for the Collapse of Cybernetics").

metaphysical and ontological issue: the possibility of an artificial, mechanical construction possessing human-like thinking capabilities. It is thus fitting to attempt approaching cybernetics from the angle of the philosophical tenets on which it was based, and the philosophical conclusions that it entailed.

CHAPTER 5

Cybernetic Tenets: Philosophical Considerations

.

PINNING DOWN THE CORE OF CYBERNETICS

Even if the proper definition of cybernetics is an issue without closure,[1] key ideas underpinning the enterprise were present right from its inception—such as circular causality, negative feedback, teleology, information, self-regulation, self-replication, and complexity. These semantically dense notions have prompted some minds to find cybernetic signatures embedded in nature as early as in pre-Socratic times. Indeed, it is claimed that even the early philosopher Heraclitus of Ephesus already had "proto-cybernetic" insights of sorts. Some arguably recognizable instances of a cybernetic flair could be mentioned:[2]

---

[1] The *American Society for Cybernetics* lists 46 serious definitions of the term. See www.asc-cybernetics.org/foundations/definitions.htm

[2] The connection between cybernetics and Heraclitus has been mentioned time and time again in academic circles, but almost at a colloquial level. A rigorous fleshing out of these possible links escapes the aim of this work—and remains to be done. The philosopher who addressed the possibility of a strong connection between them was Martin Heidegger, but he suggested that such connection is too deep for us to understand it at the present time. More on this below.

© The Author(s) 2017
A. Malapi-Nelson, *The Nature of the Machine and the Collapse of Cybernetics*, Palgrave Studies in the Future of Humanity and its Successors, DOI 10.1007/978-3-319-54517-2_5

Circular causality and feedback:

> The way up and the way down is one and the same (DK 60)[3]
> Upon those who are stepping into the same rivers different and again different waters flow (DK 12)[4]

Control:

> Thunderbolt steers all things (DK 64)[5]
> Wisdom is one thing:
> to know the Thought...
> by which all things are steered through all...(DK 41)[6]

Information:

> This world-order, the same for all...
> an ever–living fire, kindling in measures and going out in measures (DK 30)[7]
> If you have heard...not me but the Logos,
> It is wise to agree that all things are one (DK 50)[8]

> Connexions:

> ...out of every thing there can be made a unity,
> and out of this unity all things are made (DK 53)[9]
> Invisible connexion is stronger than visible (DK 54)[10]
> The real constitution of each thing is accustomed to hide itself (DK 123)[11,12]

---

[3] Marcovich 1967, p. 171

[4] Ibid., p. 206.

[5] Ibid., p. 424.

[6] Ibid., p. 449.

[7] Ibid., p. 268.

[8] Ibid., p. 113.

[9] Ibid., p. 105.

[10] Ibid., p. 36.

[11] Ibid., p. 33.

[12] The last two quotes might also be read as circular causality being more explanatory powerful than linear causality—which, of course, is itself intimately related with information.

Self-organization and equilibrium[13]:

War
    is father of all... (DK 53)[14]

Martin Heidegger, who in a later period of his life closely followed cyber-netic developments on both sides of the Atlantic, addressed the claimed connection between Heraclitus and cybernetics in his famed *Heraclitus' Seminar*.[15] Prima facie, it would seem that the notion of *steering*, funda-mental for cybernetics to the point of conforming its very etymology, and pervasive in some Heraclitean fragments, would point toward a deep shared core between the prescient pre-Socratic thinker and the novel science.[16] However, Heidegger was quick to point out that we should not confuse the Greek primal imagery of Zeus affecting nature (φύσις) using the lightning bolt as a stirring hand—a paradigmatic Heraclitean figure[17]—with a nas-cent cybernetic epiphany. In this case, for instance, it would lack the cybernetic feature of a dynamically involved systemic action, Zeus not being "stirred back" or modified in any way, by the affected φύσις. There is an absence of negative feedback, thus rendering the Heraclitean insights as not properly cybernetic. Even if there is a deep connection between Heraclitus and cybernetics, which Heidegger does not deny, it would be beyond our grasp, given our poor current knowledge.[18] Still, some persist

---

[13] Chapter 6, Section "The Homeostat: A Living Machine".

[14] Marcovich 1967, p. 145.

[15] Heidegger and Fink 1993.

[16]   At present, we reflect on the phenomenon of steering. This phenomenon has today, in the age of cybernetics, become so fundamental that it occupies and determines the whole of natural science and the behavior of humans so that it is necessary for us to gain more clarity about it (Heidegger and Fink 1993, p.12)

[17] Heraclitean fragment cited above as DK 64 (Marcovich 1967, p. 424).

[18]   I do not want to allow a misunderstanding to arise from my allusion to modern cybernetics in the course of the discussion about what steering is. Misunderstanding would arise if we restricted ourselves to what is said about steering in Frs. 64 and 41, and if we constructed a connection between Heraclitus and cybernetics. This connection between Heraclitus and cyber-netics lies much deeper hidden and is not so easy to grasp. It goes in another

in seeing a proto-cybernetic epistemology in Heraclitus—and probably not without reason.[19]

Heidegger did have a deep grasping of the cybernetic ethos, referring to it on one occasion as the "metaphysics of the atomic age".[20] In an interview he granted to the German magazine *Der Spiegel* (with the explicit condition that it was not to be made public until after his death) he referred to cybernetics in a rather dark manner. Five days after Heidegger passed away, it was accordingly published.[21] In the translation done by his lifetime friend, Father William Richardson S.J. (1920–2016), one can find a dramatic view regarding the role that cybernetics was to assume in mankind:

Heidegger: . . . The role philosophy has played up to now has been taken over by the sciences today . . . Philosophy dissolves into the individual sciences: psychology, logic, and political science.
Der Spiegel: And what takes the place of philosophy now?
Heidegger: Cybernetics.[22]

In order to attempt to further understand the gist of Heidegger's remarks, it is helpful to cradle them within the broader context of the gradual embodiment of *Machenschaft*—the gigantic "all-pervasive and totalizing 'makeability' of everything".[23] This notion, almost exclusively Heideggerian, has gained

---

direction that we could not discuss in the context of our present awareness of Heraclitus (Heidegger and Fink 1993, p. 16).

[19]   "The deepest intuitions concerning real-life complex systems date back already to Heraclitus". Hyötyniemi goes on to state three core ideas that in his view are evidently Heraclitean:

Everything changes, everything remains the same.
Everything is based on hidden tensions.
Everything is steered by all other things.

Although he does not mention this, one could relate the first one to DK12, the second one to DK 54 & 123, and the third one to DK 41 & 64 (See Hyötyniemi 2006, p. 7).

[20] Dupuy 2000, p. 90. It would seem that this is Dupuy's personal translation of a part of the Spiegel interview I reproduce below.

[21] On May 31st, 1976. The English translation appeared in 1981 (Heidegger 1981).

[22] Heidegger 1981, p. 59.

[23] Cited in MacDonald 2008, p. 177.

renewed attention in the Anglo-American world after the translation in 1996 of *Contributions to Philosophy (From Enowning)*,[24] considered by some to be Heidegger's *magnus opus* after *Being and Time*. "Machination" as it is sometimes translated, is a process bound to eventually convert everything thinkable into the buildable, and is a parallel outcome of the unfolding of technology—which in Heidegger's thought amounts to the evolution of metaphysics itself. In this later book Heidegger further explicates this heretofore somewhat cryptic notion.

> Machination is the domination of making and what is made. But in this regard one is not to think of human dealings and operating but rather the other way around; such [human activity] is only possible, in its unconditionality and exclusivity, on the basis of machination...At the same time machination contains the Christian-biblical interpretation of beings as *ens creatum*—regardless of whether this is taken in a religious or secular way.[25]

Heidegger wrote a treatise on the metaphysical foundations of logic, where he referred to Leibnitz as a major force behind "framing"[26] and machination —due to Leibnitz' ideal of a universal language based upon logic and calculus.[27] As Wiener himself did, he found in Leibnitz, the same "patron saint" for this omni-encompassing machination. However, for Heidegger the impetus of this "machinization" (as *Machenschaft* is sometimes also translated) goes well beyond the seventeen century. For him machination is nothing less than just the other side of the coin regarding Western thought itself. For Heidegger "we know too little of it, even though it dominates the history of being in western philosophy up to now, from Plato to Nietzsche".[28] It is so profoundly engrained into who we are that

> It is the double and contradictory process of "humanizing" of everything, pioneered by the Cartesian conception of the "de-humanizing", typical for an age which has been submitted to the total "machinization", in order to

---

[24] Heidegger 1999.

[25] Ibid., § 67.

[26] The word that Heidegger used is "enframing", which gives a more active role to the otherwise more passive connotation of "frame". See Heidegger 1977a, p. 20.

[27] Heidegger 1984.

[28] Heidegger 1999, § 61.

achieve the absolute grip on being, in a word, an age totally enslaved by planetary technology.[29]

Now one can be more equipped for further grasping the reason for such darkness in Heidegger's considerations regarding cybernetics. Heidegger has some rather unsettling remarks concerning the future of mankind in what pertains to the consequences of the advance of contemporary science for the human condition *vis à vis* its undisclosed cybernetic identity:

> No prophecy is necessary to recognize that the sciences now establishing themselves will soon be determined and guided by the new fundamental science which is called cybernetics ... Cybernetics transforms language into an exchange of news. The arts become regulated-regulating instruments of information. The development of philosophy into the independent sciences which, however, interdependently communicate among themselves ever more markedly, is the legitimate completion of philosophy. Philosophy is ending in the present age.[30]

Even if this impulse toward framing and machinization—which reached its zenith in cybernetics—already was with us for as long as there is metaphysics (with its own impetus for understanding, manipulating, and controlling), it was Alan Turing who let the machine finally break free. His inquiry into the nature of an algorithm, in order to properly address the nature of computability, let the door open. And even if there is consensus in that a formal and finished definition of an algorithm is still lacking,[31] it exercised a clear

---

[29] Quoted in François 2007, p. 432.

[30] Heidegger 1977b, p. 376. The quote continues:

> [Philosophy] has found its place in the scientific attitude of socially active humanity. But the fundamental characteristic of this scientific attitude is its cybernetic, that is, technological character. The need to ask about modern technology is presumably dying out to the same extent that technology more definitely characterizes and regulates the appearance of the totality of the world and the position of man in it.

[31]   When algorithms are defined rigorously in Computer Science literature (which only happens rarely), they are generally identified with abstract machines ... this does not square with our intuitions about algorithms and the way we interpret and apply results about them ... This problem of defining algorithms is mathematically challenging, as it appears that

influence in the progress of technology.[32] Probably more importantly, as in many intellectual endeavours, Turing's tangential (to the nature of a machine) but dramatic insight opened up a new realm of thinking.

> our intuitive notion is quite intricate and its correct, mathematical modeling may be quite abstract. (Moschovakis 2001, p. 919).

[32] The recognized abstract nature of an algorithm makes it ineligible for patenting:

> Determining whether the claim falls within one of the four enumerated categories of patentable subject matter recited in 35 U.S.C. 101 (i.e., process, machine, manufacture or composition of matter) does not end the analysis because claims directed to nothing more than abstract ideas (such as mathematical algorithms), natural phenomena, and laws of nature are not eligible for patent protection (United States Patent and Trademark Office 2014, § 2106, II).

In the same vein, the patenting of software (arguably a conjunction of algorithms) remains controversial, as one can witness in current news regarding the mutual lawsuits launched between global technology companies (Google, Apple, Samsung, etc.). However, certain uses of an algorithm, on entities that could qualify as "processes", are patentable. The United States Patent and Trademark Office (USPTO) designates four realms of reality that could receive patent protection:

> i. Process—an act, or a series of acts or steps ... ("A process is a mode of treatment of certain materials to produce a given result. It is an act, or a series of acts, performed upon the subject-matter to be transformed and reduced to a different state or thing." ...) (Ibid., § 2106, I).

Interestingly, the second realm patentable (a machine), is thus defined:

> ii. Machine—a concrete thing, consisting of parts, or of certain devices and combination of devices ... This includes every mechanical device or combination of mechanical powers and devices to perform some function and produce a certain effect or result. (Ibid.)

It would seem that the USPTO, not without reason, has avoided attaching intrinsic materiality to a machine ("a concrete *thing*") and has essentially linked mechanicity to an underpinning machine-structure ("this includes every mechanical device or combination of mechanical powers").

The other two areas of reality captured by the USPTO for patenting protection are iii. Manufacture and iv. Composition of matter (Ibid.).

The rediscovered awareness regarding this normally overseen nature of a machine, namely, that it did not need to be materially instantiated to be such, is what lies at the core of the cybernetic impetus. This quintessentially cybernetic tenet has been usually overseen when articulating cybernetics as a historical event. Precisely this liberation from the machine's heretofore physical constraints is what made "the machinal" amenable of instantiation in previously mechanistically unfriendly realms. The enhanced notion of *machine* substantially extended the realm of what could be tractable under a straightforwardly mechanical approach—or, in the view of the incipient cybernetic scientific community, subsumable under *science*, period.

Arguably a major contribution from Turing to the cybernetic enterprise (and eventually probably to science at large) is the insight that the existence of a mechanism *necessarily implies* the existence of a machine, namely "understanding by purely mechanical process one which could be carried out by a machine".[33] Behaving machine-like is the outcome of a mechanical structure: A machine *acts* like a machine. Indeed, pondering on this observation seems to begin to resolve some conundrums in scientific explanation. To the question of whether or not the recognition of a mechanical process in a phenomenon (natural or not) entails the recognition of a machine, one should, if switching our metaphysical dampers offline, answer in the positive.

Physical entities, previously regarded as fundamentally different from machines, are now amenable to being treated mechanically. This occurs due to the fact that physical processes canonically regarded as pertaining to living organisms only—such as self-adaptation to internal and external (environmental) change—were now found to be mechanizable. When the theoretical (mainly metaphysical) divide between the natural and the artificial collapses, one could start the endeavor of understanding and developing one theory of "control" for both animals (humans including) and machines. By "machines" we could thus refer indistinctly to both the traditional machines and to living organisms. As radical as this view may sound, it would, however, hardly qualify as radically *new*. The novel aspect of cybernetics, the one that sets it apart, is an aspect that was qualitatively distinct—and that

---

[33] Turing 1939, p. 150.

the culmination of the "Foundational Crisis of Mathematics" made possible.[34] Cybernetics now referred to immaterial entities as machines. One of the objects of study that thereby became scientifically approachable was the human mind. Animal cognition and life in general were now amenable of being treated at the same ontological and epistemological levels, as complex artificial machines: both display features of successful coping with environments—what in biology was already called "homeostasis". Recognizing that the nature of a machine is what lies at the very foundation of cybernetic thinking would put Wiener's experience with his predictor as midwifery to what was already in the making. At the very least, it qualifies re-reading cybernetics from the standpoint of its unique precepts regarding the nature of a machine.

There have been several ways of classifying the main theoretical tenets of cybernetics, each based upon its founders, its founding papers, its philosophical premises, its scientific convictions, its subsumed disciplines and even its subsequently founded disciplines.[35] This writer proposes still a new classification, based upon the conceptualization that, for our purposes, best encompasses the multifarious, multilevel theoretical signature that cybernetics came to be known for. Ross Ashby, the British cyberneticist who might have had cybernetic insights even before Wiener did, spoke of cybernetics as "the domain of 'all possible machines'".[36] Departing from this approach, one could get adequately equipped for digging into the metaphysical assumptions behind cybernetics, in order to unveil, tackle, and articulate its underlying ontology. Accordingly, this writer found appropriate to predicate of cybernetics as relying on three main theoretical pillars themselves cradled on fundamental aspects of the nature of a machine. I will proceed to spell out the cybernetic hypothesis subsuming its different foundational elements under this proposed triad.

[34] Chapter 4, Sections "The "Foundational Crisis of Mathematics" and The Response from Formalism" & "A Machinal Understanding of an Algorithm and the Material Liberation of the Machine".

[35] Dupuy 2000, ch. 2.

[36] Ashby 1956, p. 2.

MACHINES CAN BE TELEOLOGICAL

Machines are islands of order in a chaotic world. To speak of a deterministic system aiming at an ultimate purpose is not contradictory

The notion of a final cause had been cast out from mechanistic explanations due to the alleged fact that it locates "the cause after the effect", so to speak, in the natural phenomenon being studied. The cybernetic proposal of "teleological machinery" was, thus, controversial to say the least—and for some, just plain scandalous. However, this proposal was justified by retaining an important aspect of such teleological explanation, namely, that it describes a physical process as aiming toward a goal to which such process is projected or attracted, aided by corrections exerted through "negative feedback"—this last feature being characteristically present in cybernetic parlance.

It has been previously mentioned how the better understanding of the interface between man and machine came to be a crucial priority for anti-aircraft weapon development in the early 1940s. As an important aspect of this effort, the notion of a "machine" had to be meticulously dissected, in order to situate "man" as a smooth moving gear within the whole mechanical system—in fact an integral *part* of the bigger machine (the anti-aircraft gun). A salient feature of this mechanical operation is the role of feedback for the proper functioning of the weapon. Negative feedback[37] allows a machine to engage in self-correction in order to maintain stability toward a certain goal—in this case, hitting the airplane. Information had to be fed into the machine in such a way that it could reliably plot an extrapolation of the future position of the flying target, in order to shoot projectiles in the right direction at the right time. This situation was compared to a strategic movement found in nature itself, that is, when a cat hunts down a mouse. The feline would supposedly not run behind the rodent; instead, it foresees the future position of the prey, and then the hunter runs *directly* toward the precise point where the nutritious moving target would arrive.

---

[37] In contradistinction, positive feedback occurs when feedback has a reverberating effect, reinforcing the process toward the goal, without correction. An example would a radio transmitter and receiver, which picks up the signal of what it already transmitted, processing it again and broadcasting it stronger (See Chapter 2, Section "The AA-Predictor").

As we saw in Chapter 2, Norbert Wiener (after sharing his thoughts and enticing the audience at the "Cerebral Inhibition Meeting" in 1942) further articulated his claims in a 1943 article under the title "Behavior, Purpose and Teleology"—published in conjunction with the Mexican cardiologist Arturo Rosenblueth and the MIT engineer Julian Bigelow. Publishing this article introduced to a larger audience the notion of a teleological machine: a machine whose operation toward a set goal is kept in check by negative feedback, without human intervention. Negative feedback in a physical process had as its aim to regulate the mechanical operation in such a manner that the disturbances that would direct the processes away from its goal are corrected—checked and put back on track. A paradigmatic example of this dynamical interaction occurs in the famed steam governor mentioned in the first chapter. The three cybernetic authors asserted in this article that a self-regulated anti-aircraft gun machine, a self-guided torpedo, and an airplane's auto-pilot, are all examples of teleological machinery.

Although the article was primarily aimed to a scientific audience, these statements expectably caught the attention of the philosophical community. Indeed, since the idea of a "final cause" always was a problematic issue in philosophy and science, the notion of a teleological machine attracted the interest of some philosophers of science—with the expectable sharp criticism. Attacks against the usage of the term "teleological" were soon targeting the cybernetic authors in a typically severe Anglo-Analytic manner. In 1950, the then young philosopher Richard Taylor[38] vigorously attacked the cyberneticists' choice of "teleology" as a wishful scientific neologism, claiming that it shared only a superficial resemblance with the classical term used in philosophy from times immemorial. But precisely keeping this term was indeed consciously desired by the cyberneticists, arguing that, with some modification,[39] it conformed a quintessential aspect of the whole cybernetic stance. They would not want to give it up. Disregarding this, Taylor claimed that, in principle, a torpedo being pulled by a ship by means of a wire or rope attached to its nose could also qualify as a "teleological machine", since its "aim" (eventually reaching and hitting the vessel) would also be accomplished—and indeed even by

---

[38] Taylor 1950a.

[39] For example, foregoing the notion of a final cause, but retaining the notion of a pattern of behavior that aims toward a goal.

means of correcting possible disturbances in its trajectory. He trivialized Wiener's view of teleological behavior.

Rebuttals came from both sides.[40] Taylor's criticism was solid and well argued. Nevertheless, Wiener and Rosenblueth firmly stood by their ideas. Probably their strongest point against Taylor was that he was still operating within a notion of physics that could be understood as thoroughly Newtonian—the so-called "philosopher's physics". Causality in complex systems occurs in non-linear ways, and the cyberneticians were referring to those. Also, Wiener and Rosenblueth hinted at some of the problems raised by Taylor as being purely "verbal" in nature, lacking any substance. Finally, they just dismissed some of the philosopher's concerns as simply not worth addressing. Wiener published in 1945 a piece on the "The Roles of Models in Science"[41] to justify his choice of retaining the somewhat modified notion of teleology, so that it could be applied to mechanical systems—and he referred the reviewer (Taylor) to this article in his response to the critical piece. It is worth noting that of the 10 conferences held by the Macy foundation, a professional philosopher was invited only to the first one—Filmer Northrop (1893–1992). Interestingly, however, most cyberneticians pursued formal studies in classical philosophy for some time earlier in their academic lives— particularly so in the cases of Norbert Wiener and Warren McCulloch.

There was, however, another episode that would point out to a later deep rift in scientific worldviews among cyberneticians regarding the idea of behavior-with-an-aim. This rift might have partly spelled out the implosion of the enterprise itself at a theoretical level. This could indeed be one of the first instances where a new paradigm of science, which is reportedly still advancing well in the twenty-first century,[42] trumpeted across the legacy of a previous one, bringing the latter into oblivion. Cybernetics might have inadvertently been one of its first victims. This assertion requires some explanatory context.

Just as Wiener's ideas were already known in relevant circles before the publication of his book in 1948, a younger mathematician and engineer was, around the same time, beginning to gain notoriety. Since his years as an MIT electrical engineering graduate student, Claude Shannon

[40] Wiener 1950; Taylor 1950b.

[41] Rosenblueth and Wiener 1945.

[42] Prigogine and Stengers 1984.

(1916–2001) was already interested in Wiener's work, knowing that the latter was doing important research at the same institution. While pursuing a Master's degree, Shannon became an assistant for Vannevar Bush' second differential analyzer previously mentioned.[43] Once he landed a permanent position at Bell Laboratories, war-related work put both Wiener and him in direct contact, since Wiener had to deal with the Bell Company on a frequent basis (both Bell Labs and MIT were funded under the referred above subsection D-2). Shannon started to pay Wiener frequent visits at his research lab, seeking inspiration and advice on relevant topics. These tours became so frequent indeed, that at a certain point—and aware of the closely related research occurring at Bell Labs—Wiener begun to be concerned about the possibility of having his ideas siphoned out without his due acknowledgement.[44] In fact, the same year that Wiener published *Cybernetics*, Shannon published the famed "A Mathematical Theory of Communication"—a long article delivered in two installments, soon acquiring legendary status within engineering circles.[45]

Warren Weaver, still supervising the military-related research of section D-2,[46] endorsed the paper by later expanding it into a joint book—coauthored by Shannon and himself—and given the more ambitious title *The Mathematical Theory of Communication*.[47] The monograph included Shannon's paper and Weaver's simplified explanation and assessment of the theory's foreseen import, taking advantage of the latter's gift for conveying otherwise cryptic concepts—both as a people's person and in writing. The paper acknowledged very clearly Wiener's influence, and

---

[43] See Chapter 2, Section "The Pre-meetings and Founding Articles". Shannon's Master's thesis was an application of Boolean algebra to the workings of such machine. He went on to pursue a relatively quick Ph.D. in mathematics, also at MIT, proposing an algebra applicable to Mendelian genetics. It was right after earning this degree that he began working for Bell Laboratories.

[44] Conway and Siegelman 2005, p.186.

[45] Shannon 1948.

[46] Although Warren Weaver was not directly immersed in the cybernetic project, partly due to his role as science administrator (rather than practitioner), he was by proxy firmly related to the circle. However, he was directly involved with some posterior developments that came out of cybernetics. More on this in Chapter 9, Section "Cybernetics 2.0: The Nano-Bio-Info-Cogno Convergence".

[47] Weaver 1949.

Wiener was thankful and respectful of such homage.[48] Shannon carried on Wiener's insight on the digitality of information, asserting logarithm at base 2 at the core of the general equation of information. In fact, even if Shannon's Bell Labs colleague John Tukey first coined the term "bit" for "binary digit",[49] Shannon established the term as a core part of information theory's structure.[50] However, Shannon's proposal had, as Wiener publicly praised, original features of its own.

As advanced above,[51] the paper started putting forward (or "clarifying", in Shannon's view), the controversial claim that information has nothing to do with meaning—or rather, that the problem of semantics should be bracketed as irrelevant to the fundamental problem of communication— namely, "that of reproducing at one point either exactly or approximately a message selected at another point".[52] As shown in the first chapter, when referring to the 8th Macy conference, several people were concerned about this move away from the "significance" of the message conveyed—among those being Donald Mackay and the editor-in-chief, Heinz von Foerster. However, much of this unease might had been partly triggered by a misunderstanding fostered by the intense enthusiasm of cybernetics itself.[53]

---

[48] Later on, Wiener would become more assertive, even protective, of the role that he himself played in developing "Shannon's theory". Cf. Conway and Siegelman 2005, pp. 187–188.

[49] Shannon 1948, p. 380.

[50] Boden 2006, pp. 204–205; Hayles 1999, pp. 50–57.

[51] Chapter 2, Section "The Macy Conferences".

[52] Shannon 1948, p. 379.

[53] A recorded conversation between Claude Shannon and his wife Betty, toward the end of his life, seems to point in that direction:

Betty: In the first place, you called it a theory of communication . . . You didn't call it a theory of information.

Claude: Yes, I thought that communication is a matter of getting bits from here to here, whether they're part of the Bible or just which way a coin is tossed . . .

Betty: It bothered you along the way a few times but by that time it was out of your hands.

Claude: The theory has to do with just getting bits from here to here . . . That's the communication part of it, what the communication engineers were trying to do. Information, where you attach meaning to it, that's the next thing, that's a step

In any case, due to its simplifying character, more suitable for engineering purposes, Shannon's take prevailed as the approach of choice.[54]

An interesting shift away from Wiener's views, however, occurred in Shannon's paper. Wiener's views on entropy, as previously pointed out,[55] largely came to him out of the realization of deep correlations found between on the one hand his early works on chaos and thermodynamics, and on the other the nature of the message occurring between internal structures and their environment in surviving systems. This context led Wiener to equalize information with negative entropy. Information provided organization to a reality that inherently tends toward chaos. But Shannon stated that information was just the opposite: that information was entropy itself—and not negative entropy. This reversal carries relevance since it would come back later in disguise with important consequences for the cybernetic enterprise.

As Weaver explained Shannon's view of information, "information is associated with the amount of freedom of choice we have in constructing messages".[56] If the message conveyed happens to bring to the receptor what the receptor already has, then little information has been conveyed. If, on the other hand, there is a high degree of uncertainty regarding the possibility of reconstructing the message at the receptor's end, then one can talk of a greater amount of information at stake. So it would seem that information would amount to being a function of the amount of randomness in the conveyance of a message—or as Weaver would put it, "a measure of one's freedom of choice in selecting a message".[57] Indeed, Weaver would cleverly capture the case where a good degree of order is present, and the amount of choice is poor, saying that this "situation is highly organized, it is not characterized by a large degree of randomness or of choice, the information (or the entropy) is low".[58]

beyond, and that isn't a concern of the engineer, although it's interesting to talk about (Conway and Siegelman 2005, pp. 189–190).

[54] Not without some still clinging for a while longer to Mackay's more cumbersome (for engineering) approach (Hayles 1999, p. 56).

[55] Chapter 2, Section "The Macy Conferences".

[56] Weaver 1949, p. 12.

[57] Weaver 1949, p. 9.

[58] Ibid., p. 13.

In contrast, Wiener stopped short of identifying entropy with "evil": we are after all destined to turn to ashes at the end, due to reality's tendency toward chaos and disorganization. Should we thus consider Shannon's approach as inimical to that ultimate challenger of entropy, "life" (that most excellent embodiment of information)? That would not seem to be the case, given Shannon's keen interest in synthetically recreating instances of living systems, as portrayed in the construction of his famous robotic "maze-solving rats".[59] It would also be relevant to point out that the scientific and intellectual environment regarding the physical and its semantic implications for chaos where about to begin undergoing revision. Entropy, or the randomness present in a physical environment started to be seen as a fruitful context for complexity to arise. Physical systems, organic or not, had to engage in a certain degree of self-organization in order to be able to cope with that changing physical environment. Lack of flexibility would entail destruction—and death.[60]

Wieners' view on entropy as contrary to life and organization largely comes from the observation of an isolated system's behavior under the 2nd law of thermodynamics. Since energy is not created or destroyed, in an exchange of it something has to give, and that would be the order (information) that maintains the system together.[61] This inverse relational status between entropy and life was sanctioned by Schrödinger[62]—which

---

[59] Hayles 1999, pp. 63–65.

[60] The received view of entropy as the spiraling into chaos and oblivion inherent to nature gets questioned as the default state of affairs of things (Prigogine and Stengers 1984). More on this below.

[61] "Every process, event, happening—call it what you will; in a word, everything that is going on in Nature means an increase of the entropy of the part of the world where it is going on ... An isolated system or a system in a uniform environment ... increases its entropy and more or less rapidly approaches the inert state of maximum entropy. We now recognize this fundamental law of physics to be just the natural tendency of things to approach the chaotic state ... " (Schrödinger 2012, p. 71).

[62] "[A] living organism continually increases its entropy—or, as you may say, produces positive entropy—and thus tends to approach the dangerous state of maximum entropy, which is of death. It can only keep aloof from it, i.e. alive, by continually drawing from its environment negative entropy ... What an organism feeds upon is negative entropy. Or, to put it less paradoxically, the essential thing in metabolism is that the organism succeeds in freeing itself from all the

Wiener approved. Once the environment is factored into the picture, it would seem that the emergence of complexity (partly due to random changes in order to cope) is now the friendly context for life to emerge. Indeed, it seems as if self-organization would need such an explosion of variables to successfully match a changing milieu and reach homeostasis.

Claude Shannon, in that respect, advanced what was later given empirical plausibility by the 1977 Nobel Prize in chemistry Ilya Prigogine (1917–2003) for his study of self-organization and thermodynamics. It would seem that order in nature (and life itself) are necessary processes of reciprocal coping balances that are actually bound to happen—*contra* the traditional view regarding the consequences of entropy...[63]

> Matter near equilibrium behaves in a "repetitive" way. On the other hand, far from equilibrium there appears a variety of mechanisms corresponding to the possibility of occurrence of various types of dissipative structures. For example, far from equilibrium we may witness the appearance of chemical clocks, chemical reactions which behave in a coherent, rhythmical fashion. We may also have processes of self-organization leading to nonhomogeneous structures to nonequilibrium crystals.[64]

The consequences of this surreptitious semantic change regarding information as a fundamental element of cybernetics might have entailed a deep displacement in the ontological commitment underlying the project. I will come back to these effects in the chapter regarding the work of William Ross Ashby.[65]

---

entropy it cannot help producing while alive...Thus the device by which an organism maintains itself stationary at a fairly high level of orderliness (= fairly low level of entropy) really consists [in] continually sucking orderliness from its environment." (Ibid.)

[63] "At all levels, be it the level of macroscopic physics, the level of fluctuations, or the microscopic level, *nonequilibrium is the source of order. Nonequilibrium brings "order out of chaos"* (Prigogine and Stengers 1984, pp. 286–287. Italics original).

[64] Prigogine and Stengers 1984, p. 13.

[65] Chapter 6.

MACHINES CAN BE IMMATERIAL

In virtue of this feature, they can be "extended" to encompass, beyond the organic realm (life), non-physical entities (mind) as well

It has been mentioned that McCulloch was tinkering with the notion of idealized neurons, in connection with the premise that there should be a logical way of transducing cognitive processes—so that we can have a glimpse of hope for relating the exactness of mathematics with the chaos of thinking. Indeed, McCulloch already had for a while a nagging insight which suggested that the nervous system was nothing but the "wetware" instantiation of a higher, more universal, and abstract logical machine. In fact, his artificial neurons would stand for the logical gateways existent thanks to Boolean logic ("and", "or", etc.). It is in this context that Turing's observations regarding an "ideal" machine importantly shaped McCulloch's own insightful musings. He recalls that

> it was not until I saw Turing's paper that I began to get going the right way around, and with Pitts' help formulated the required logical calculus...The important thing was, for us, that we had to take a logic and subscript it for the time of the occurrence of a signal (which is, if you will, no more than a proposition on the move). This was needed in order to construct theory enough to be able to state how a nervous system could do anything. The delightful thing is that the very simplest set of appropriate assumptions is sufficient to show that a nervous system can compute any computable number.[66]

This model, which re-instantiated a pristine logical platform into an object of reality, thus "purifying" the latter, was a major impulse for cybernetics, which was now holding a mechanical explanation for the mind's hitherto scientifically intractable operations. The article turned out to be a theoretical pillar not only for cybernetics, but also for the start-up of the field of computer science. John von Neumann's famed *First Draft of a Report on the EDVAC*[67] used, to the puzzlement of the engineers hired to build these early computers, McCulloch's vocabulary and symbols of "neurons" throughout his paper. He was open about having written the draft based

[66] Von Neumann 1951, p. 32.
[67] Von Neumann 1945.

upon McCulloch and Pitts' proposal.[68] Von Neumann's paper served as a blueprint for the construction of one of the first "stored-program" computers (the EDVAC[69]) to be delivered, as it was expected given the urgencies of the time, for ballistics research at the Aberdeen Proving Ground.[70] The model can be explained in the following way.

For instance, in the case of an organism (e.g., a bird) "deciding" whether or not eating an object (e.g. a blueberry), something like the ensuing procedure would take place. The animal's neuron would receive two possible inputs, one regarding shape (roundness) and the other regarding color (purple). When the bird's biological detector perceives "roundness", it would fire a signal (symbolized by a 1). If it does not—if the object is square—then there would be no firing (symbolized by a 0). By the same token, if the detector perceives "purple", then 1 would occur—if it perceives blue, then 0 would happen. The situation that we have thereafter is that when the bird perceives a blueberry, a threshold would be formed in the ideal neuron. This threshold would be represented by the number 2, formed as a result of detectors firing due to the perception of both roundness (1) and purpleness (2): $1 + 1 = 2$. If this threshold is reached, the neuron will this time itself fire a response, represented by 1 if the command is "eat", and by 0 if it is "don't eat". Only if a threshold that is equal or more than 2 is reached, then positive command will occur. Since both the roundness and purpleness of a blueberry causes the detectors to send overall signals that reach the 2 threshold, then a 1 signal will be fired and the bird will eat it. If on the other hand, the bird sees a golf ball, there would be roundness (1) but no purpleness (0), forming a total output of just 1 $(1 + 0 = 1)$. Since the threshold requires a minimum of 2 for the neuron to fire "eat" (1), the bird would not eat. A similar situation would happen with a violet, where there is purpleness (1) but not roundness (0), and so the total output would still be below the minimum required threshold of 2 $(1 + 0 = 1)$.

[68] Von Neumann 1945, p. 5.

[69] Electronic Discrete Variable Automatic Computer.

[70] Situation which, as we will see below, signaled the different orientation that Americans, in contrast with its British Allies, had regarding the war efforts—which reverberated in different ways of practicing cybernetic science. More on this in the next Section "The Backlash: Unforeseen Consequences of a Behavior-based Ontology".

Still in the case of a hotdog, lacking this food both roundness and purpleness, the outcome will fall further below the minimum required to reach the 2 threshold—since $0 + 0 = 0$. And so on and so forth.[71] Expectably, McCulloch and Pitts' networks get quickly more complicated once more features are added into the operation.

Part of the fame of this article resides in its almost epic difficulty. John von Neumann was quick in pointing out that its cryptic nature was in great part due to the usage of Rudolph Carnap's *sui generis* and somewhat convoluted symbolic nomenclature. This unfortunate choice—in the eyes of von Neumann—was probably unavoidable anyway, since Carnap was one of Pitts' admired mentors. Despite its inaccessibility (which partly almost cost it its publication), the paper became a theoretical bedrock for cybernetics. It was not, however, devoid of problems.

The complexity that is rapidly attained after more elements enter the cognitive picture turned out to be particularly problematic in a two-fold manner. On the one hand, the networks, even if complex, were still considered too simple for occurrences of higher cognition. On the other, even if immensely more complex cognitive processes were to be mapped, the constructability of the model would get severely compromised. Since cybernetics had as a signature feature the buildability of its models,[72] this also brought about theoretical unease (in fact, there were later developments of the model, largely side-stepping the constructability mandate—e.g., the perceptron[73]—thus to some extent betraying the cybernetic mandate of material grounding).

Pointing out the importance of this paper for the theoretical genesis of computer science is indeed relevant for a better understanding of what later occurred to the cybernetic enterprise. Right from the start, von Neumann, who certainly saw its usefulness—even built an impressive physical device out of it—insightfully saw its shortcomings as well. These weaknesses became evident for him when the model was applied to what was supposed to understand in the first place: the mind. When the neatly abstract and disembodied machine is instantiated back into materiality, a remodeling of the studied object seems to occur, severely complicating it

---

[71] Marsalli 2006.

[72] See the following section. Also Pickering 2005.

[73] Chapter 3, Section "Traditional Explanations for the Collapse of Cybernetics".

right into intractability, evincing deep theoretical problems in the model. Thus it is from insights into the feasibility of the McCulloch and Pitts' model into realms other than computer machines, that the most severe criticisms against the core of cybernetics would be launched.[74]

MACHINES ARE EMBODIED THEORIES

Physical constructability should act as theory grounding. Theoretical investigation ought to be constrained by a concrete empirical buildability

Probably the most famously peculiar characteristic of the cybernetic movement was its emphasis on the necessary constructability of whatever was proposed—instantiated in cybernetics' legendary autonomous robots. Certainly this leniency toward the engineering aspect of a theory has older roots.[75] Constructability turns out to be the strongest sign of a model being truthful and reliable: knowledge entails buildability. True epistemic assets entail engineering success, our model enjoying an equilateral isomorphism with the modeled.[76]

It is relevant to point out that the cybernetic collaboration coming from Britain had this distinctive feature, which hitherto lacked, at least in such measure, in the American side of cybernetic research: machine building as theory testing.[77] In fact, it could be maintained that if the English were not as actively engaged as Americans in cybernetic research (as both McCulloch and Wiener asserted after visiting their English counterparts),[78] the British definitely compensated for that alleged lack with their emphasis on machine building. This aspect of British cybernetics

---

[74] The problematic epistemological strategy that the construction of McCulloch and Pitts' networks entailed will be further addressed in the chapter pertaining to the contribution of John von Neumann (7).

[75] More on this in Chapter 9, Section "Viconian Constructability as a Criterion of Truth".

[76] The psychoanalytic assessment of what supposedly lurks behind the mostly male-filled area of Artificial Intelligence, the so-called "womb envy" syndrome, could in fact be subsumed under this more primogenial philosophical rubric referred to in the previous note. See McCorduck 2004, ch. 5.

[77] This likely is the legacy of experimentation of the Royal Society since Bacon, Boyle and Newton. I thank Jagdish Hattiangadi for pointing this out.

[78] See Chapter 3, Section "The Ratio Club".

might have been partly the result of the particular orientation of their military research, due to their own urgencies of war: less theoretical investigation and more quickly built defensive gadgetry. Such a circumstance would not only put a constraint in terms of time and secrecy (which the Americans also had), but also foster a hands-on approach to scientific research. In addition, since cybernetics did not have in the UK the official financial backing that it enjoyed in the USA, much of the British cybernetic developments took place as side jobs, even as hobbies, being performed in the scientists' kitchens and garages during their spare time.[79]

The physical embodiment of an otherwise non-material machine, not only as a heuristic tool but also as the ultimate ontology of a living object, was an idea already present in a premonitory way in early twentieth-century British psychology. Both the Ratio Club and a thinker who never made it to his first meeting due to an untimely death, become here distinctly relevant. Kenneth Craik (1914–1945) was a young philosopher with an intense interest in the workings of the mind and a confessed passion for machinery. After studying philosophy, he began pursuing a Ph.D. in psychology. In 1943, five years before Wiener's *Cybernetics*, he published a book entitled *The Nature of Explanation*. The thesis in this book had profound implications for cybernetics on both sides of the Atlantic.

Craik advanced the idea that one of the most startling capacities of the mind was the possibility of putting together prediction models about the state of affairs of reality. That is largely what thinking would amount to. In principle, Craik stated, a machine should be capable of doing the same: be able to predict situations in reality. In fact, the mind was itself, a type of machine capable of constructing models of the world. This manner of understanding reality, by means of actual model-building, was called the "synthetic method".[80] And this idea began to make a profound impression in those who would later become British cyberneticists. Emphasis in robot making as a heuristic tester for psychological theories was his explanatory signature. However, just two years after the publication of his book, and just before the first Ratio Club meeting, he was killed in a bicycle traffic accident—at only 31 years of age.

---

[79] Ibid. Also Pickering 2010, p. 10.
[80] Cordeschi 2008a.

It has been previously pointed out that the Ratio Club, in an important way was marked by this legacy.[81] In fact, the group, which deeply lamented this loss, considered naming itself the "Craik Club". Craik's emphasis in his "synthetic method" as a way to understand physical phenomena (i.e., gadget building) allegedly set the British cyberneticians apart from their American counterparts—mainly focused in mathematical expansions of what could be physicalized by means of extending the notion of what can be understood mechanistically. Also, the social implications of cybernetic thinking, which allegedly gave the American side a strong footing in terms of mainstream attention and founding, was entirely avoided by the English group. Indeed, Grey Walter, who later became a "media personality" of sorts due to his artificial tortoises, not only openly acknowledged receiving inspiration from Craik, but furthermore, characterized the American cybernetic movement as embodying Craik's ideas—minus the fun.[82]

Having in consideration that Ross Ashby himself—who was eager to convey the plausibility of his ideas by means of showing, in the metal, constructed devices—kept active correspondence with Craik, one would be tempted to agree with such remark.[83] In fact, Craik was already publishing in 1943 what would provide important cybernetic insights later—the same year when Wiener was just publishing his foundational paper with Rosenblueth and Bigelow, and full five years before the publication of his book in 1948. These timeline considerations could point out to a prescient philosophy that could concede to Craik the position of a cybernetic pioneer. However, in all fairness to American cybernetics, one would have to also consider the machines built in the USA, and recognize that they did not play a small role in the development of cybernetic thinking. A list of relevantly built cybernetic machines, both British and American, would be:

– Ross Ashby's Homeostat: a machine that will have substantial implications for cybernetic thinking.[84]

---

[81] Chapter 3, Section "The Ratio Club".

[82] " ... thinking on very much the same lines as Kenneth Craik did, but with much less sparkle and humour" (Husbands and Holland 2008, p. 14).

[83] Husbands & Holland (ibid.) share such opinion.

[84] Referred to in Chapter 6, Section "The Homeostat: A Living Machine" of this work.

- Grey Walter's "tortoises": phototropic, self-charging robots that launched cybernetics to popular culture.[85]
- Wiener's (and Bigelow's) Predictor: the machine that inspired the 1943 foundational article.[86]
- Von Neumann's computer: mapped upon McCulloch and Pitts' article, and then the source of further cybernetic thinking.[87]
- Claude Shannon's "rats": maze-solving small mobile robots.[88]

The first two machines were British. The remaining three were American. To recall, an important aspect of the operation of anti-aircraft weaponry that jumpstarted the cybernetic movement in America was the need to make the calculations for future positions of the target aircraft, tractable in a timely way, for the expectably urgent reasons that characterize war times. If done by hand, these procedures could have taken months. There was a pressing need to develop a mechanized way to *compute* these numbers. This set America on a different path from Britain concerning the orienta-tion and purpose of their computing research. After Turing's contribution with his *On Computable Numbers* to the understanding of the notion of a machine, by means of spelling out what an algorithm stands for, John von Neumann, almost in parallel with the British allies, embarked upon the goal of constructing a machine capable of doing algorithmic computations —among other tasks. The British were instead focused on cracking the coded German communications. Meanwhile, the American interest in computer building morphed into von Neumann's involvement with the EDVAC project, a computing machine to be delivered to the Ballistics Research Laboratory of the Aberdeen Proving Ground in Maryland—the same facility where Norbert Wiener was doing ballistics research, applying his studies of Brownian motions.

As pointed out in Chapter 2, it has to be said for the American side of cybernetics regarding the anchoring of theoretical advancement in physi-cal substrata, that Wiener was right from the beginning of his projects vehemently pushing for required engineering practical skills for his

---

[85] Pickering 2010, pp. 37–54.

[86] Referred to in Chapter 2, Section "The AA-Predictor" of this work.

[87] Referred to above and in Chapter 7, Section "John von Neumann's Appropriation of McCulloch and Pitts' Networks".

[88] Hayles 1999, pp. 63–65.

working team—given his past own misfortunes with the hands-on aspects of his theoretical musings.[89] In fact, he politely scolded von Neumann at some point for not having sufficiently performed a transition from designing computing architecture to immersion into control theory.[90] He looked for scientists with hands-on knowledge in at least basic electrical engineering—even as a hobby (e.g., radio construction). This type of approach underpinned Wiener's whole project of the Anti-Aircraft Predictor, as also seen by his enthusiastic support of Julian Bigelow's ingenious simulator. Hence, there was a clear and unambiguous emphasis in cradling mathematics and physics in the engineering-conscious, nitty-gritty sensitive, aspect of machinery. In Wiener's mind, proper modeling was of utmost importance in the advancement of scientific knowledge, as both him and Rosenblueth emphasized in their 1945 paper on "The Role of Models in Science",[91]—stating the peculiar claim that "the best material model for a cat is another, or preferably the same cat".[92]

It is suggested that the cybernetic emphasis on this useful theoretical constraint—the need to ground it in physical operability—is what made cybernetics a qualitatively different proposal in the history of science. It allegedly shifted the usual theoretical focus from being epistemologically based to one founded on the ontology of objects.[93] Indeed, testing how a certain philosophical or mathematical insight would work-out by means of observing the coping of a machine (metallic or organic) with the environment, fundamentally mattered. This approach in fact went well beyond a mere methodological choice for scientific research. As seen above, Wiener regarded the reality of a "machine" as "islands of hope", as pockets of intention and will, in the midst of the deadly indifference of entropy—the ever-present dark reminder that reality spirals down into oblivion.[94]

[89] As seen on Chapter 4, Section "A Machinal Understanding of an Algorithm and the Material Liberation of the Machine" of this work. Also Galison 1994, pp. 233–252.

[90] Galison 1994, p. 249.

[91] Rosenblueth and Wiener 1945.

[92] Rosenblueth and Wiener 1945, p. 320.

[93] Pickering 2010, ch. 2.

[94] Chapter 2, Section "The Macy Conferences" and Chapter 3, Section "Norbert Wiener's *Cybernetics*".

Considering the importance that machine building had for the cybernetic ethos, there is something to be said regarding the way in which these instantiations affected its own theoretical cores. The next two chapters will attempt to show the influence that the relevant constructed machinery had on cybernetics' evolution itself. One machine will be British, by Ross Ashby, and the other(s)[95] American, by von Neumann. This machinery provoked, on the one hand, the unveiling of problematic commitments that were not in the open when they still remained solely in the realm of theory. On the other, it triggered retreat into realms rejected by cybernetics in the first place as part of its fundamental mandate, going against quintessentially cybernetic standpoints.

[95] One physically embodied, the other immaterial.

CHAPTER 6

Extending the Scope of a Machine Ontology

William Ross Ashby's Nature-Machine Equalization

If one is to make a philosophical survey of what the notion of *machine* stands for, one will be quickly surprised by the fact that there is little written on the topic per se—"surprised" because philosophers and scientists have been extensively using its derivatives (mechanistic, mechanical, machinal, etc.) for at least three centuries by now. Indeed, upon tracing back the history of modern science, one cannot find a rigorous attempt at a definition of a machine.[1]

---

[1] The closest attempt at a definition found by this writer is Heinrich Hertz'— although he does not dwell on the nature of a machine. Hertz briefly states the following, when exposing the main elements of his *Principles of Mechanics*: "A system whose masses are considered vanishingly small in comparison with the masses of the systemswith which it is coupled, is called a machine" (Hertz 1899, § 531).

© The Author(s) 2017
A. Malapi-Nelson, *The Nature of the Machine and the Collapse of Cybernetics*, Palgrave Studies in the Future of Humanity and its Successors, DOI 10.1007/978-3-319-54517-2_6

There have been good studies on the notions of automaton,[2] artifact,[3] and even gadget,[4] which go all the way back to Ancient Greece.[5] However, these refer, once again, to derivatives of the more general and substantial notion of a machine, and not to the object, or the idea of the object, in itself. This reason alone (namely, the lack of a historical factum pointing toward a rigorous treatment of the idea of a machine) would already justify addressing the philosophical underpinnings of what a machine stands for. But there is a more relevant, compelling reason for squarely tackling the idea of what a machine is. Ashby attempted to define cybernetics as the theory of all possible machines.[6] This machine-based approach to the understanding of cybernetics gained strength after it was gradually realized that for under-standing animal (man) and machine under the same mechanistic laws,[7] a clarification of what a "machine" actually meant was desirable—not in an altogether different fashion as to what occurred a couple of decades earlier, concerning the need for articulating the notion of an algorithm, in order to realize what can and cannot be computable.[8]

The scientist who somewhat mysteriously asserted this preeminence— of the machine as lying at the hard core of cybernetics—went by the name

---

[2] Chapuis and Gelis 1928.

[3] Margolis and Laurence 2007; Read 2009.

[4] Berryman 2003.

[5] Ancient Greeks report the construction of mechanical marvels with unique func-tions. Their attraction to mechanical devices was mostly directed toward the aspect of clever and useful tools. Archimedes' inventions could be brought up as illus-trative examples. Medieval times witnessed an idiosyncratic infatuation with machinery, epitomized in the construction of incipient logical machines. Ramon Lull's rotating concentric circles and *volvelles* in general could be representative instances. Modern science was the locus for a renewed interest in machines, this time chiefly taking in consideration the heuristic capacities that they allegedly carry in terms of explanatory power. Rene Descartes and Leonardo Da Vinci could be pointed out as prime examples of thinkers who shared a strong interest in machinery.

[6] Quote above (Ashby 1956, p. 2).

[7] Implicit in the very subtitle of Norbert Wiener's *Cybernetics: Or, the Theory of Control in Animals and Machines* (Wiener 1948a).

[8] Chapter 4, Section "A Machinal Understanding of an Algorithm and the Material Liberation of the Machine".

of William Ross Ashby. This British medical doctor and psychiatrist was a founding member of the Ratio Club: the famed "study group" that stood as the (humbler) English parallel to the Macy gatherings in the USA, and which included Alan Turing and Grey Walter as regular members.[9] Ashby's theoretical developments were imbued, pretty much right from the beginning, with the characteristic cybernetic signature. Already in 1940, a full three years before both foundational articles by Wiener, McCulloch and company saw day light,[10] Ashby published a distinctly cybernetic article that would go on to serve as the cornerstone of his own subsequent plethora of ideas.[11]

Ashby confessed that for several years an apparently insurmountable difficulty regarding the possibility of understanding a living organism as a machine haunted him. There seemed in place a default view of an utter incompatibility between the notions of a machine and, say, a brain—due to the notorious adaptive capacities of the latter, allegedly inherently lacking in the former. Candidly referring to himself in the 3rd person, he shared:

> The author has been concerned for some years with attempting to reconcile the fact of the occurrence of this type of adaptation with the usual hypothesis that the brain is a machine, i.e., a physico-chemical system. In the opinion of some, there is a fundamental impossibility that any isolated machine could show behaviour of this type.[12]

---

[9] Chapter 3, Section "The Ratio Club".

[10] Rosenblueth et al. 1943; McCulloch and Pitts 1943.

[11] Ashby 1940.

[12] Ashby 1947a, p. 44. Ashby was more specific regarding this problematic duality five years later, at the beginning of the 1952—first edition—of his *Design for a Brain* book:

> On the one hand the physiologists have shown in a variety of ways how closely the brain resembles a machine: in its dependence on chemical reactions, in its dependence on the integrity of anatomical paths, and the precision and determinateness with which its component parts act on one another. On the other hand, the psychologists and biologists have confirmed with full objectivity the layman's conviction that the living organisms behaves typically in a purposeful and adaptive way. These two characteristics of the brain's behavior have proved difficult to reconcile, and some workers have gone so far as to declare them incompatible (Ashby 1952a, p. 1).

Ashby thought that what may lie at the crux of this claimed incompatibility might be a poor understanding of the entity to which the nervous system is being compared. He asserted that "[i]t seemed possible, however, that a more elaborate study of the essentials of 'machines' might show possibilities hitherto overlooked".[13] Accordingly, he proceeded to gradually break down the task at hand—namely, to find out what a machine is—into pieces, carefully re-building the notion, along with the isomorphic characteristics allegedly shared with the nervous system, in a meticulous and sequential fashion. His suggestion for a "more elaborate study" took him almost a decade.

As stated above, Ashby's cybernetic argumentation started in 1940. In an article entitled "Adaptiveness and Equilibrium", he began by addressing the very feature that supposedly sets the nervous system irreconcilably apart from a machine, advanced in its title: *adaptiveness*. He proposed equalizing it with the familiar and recurring physical phenomenon of *equilibrium*. This should, for him, provide an objective approach to adaptiveness, void of subjective appreciations and metaphysical preconceptions. For this, he began scrutinizing the said notion of equilibrium, incrementally building up on top of some basic propositions.

For heuristic purposes, Ashby encouraged the reader to imagine three physical objects, each showing a distinct type of equilibrium: a *cube* lying on one of its sides (stable), a non-moving *sphere* (neutral) and an inverted *cone* resting on its tip (unstable). If we try to tilt the cube, it will resist before it reaches 45 degrees; after that, it will topple over. If we push the sphere, it will roll. If we touch the inverted cone, it will fall on its side. In regards to the resulting distribution of forces, one could say that in the cube's case the disturbing force acts against the resultant force; in the case of the sphere the resultant force *is* the disturbing force (there is only disturbing force); and in the case of the inverted cone the resultant force acts along with the disturbing force.

This quantitative approach to what equilibrium stands for suggests two important features: (1) Equilibrium does not imply immobility: a perfect pendulum, with a frictionless hinge, stands for a system in equilibrium. (2) Equilibrium is inherently dynamic: if the three objects alluded above were unaffected by any disturbing force, ascribing any

---

[13] Ashby 1947a, p.44.

type of equilibrium present would hardly make sense—one could say at most that Newton's First Law is being witnessed.[14]

At this point, Ashby was ready to more clearly specify what equilibrium is, advancing that "a variable is in stable equilibrium if, when it is disturbed, reactive forces are set up which act back on the variable so as to *oppose* the initial disturbance. If they go with it then the variable is in unstable equilibrium".[15] Once he accomplished this clarification, Ashby made an interesting move. He proposed that every system in equilibrium implies a circuit. How could this be so? The system of opposing forces reveals a circuital flow: a variable that is affected by the disturbing force affects back the disturbing force itself. Graphically, it could be displayed as something like this:

$$X_1 \; \underset{\longleftarrow}{\overrightarrow{}} \; X_2$$

Moreover, Ashby further stated that the reverse is also true: that the existence of a circuit entails an existing equilibrium! And this is so because the flow of disturbance to the variables also denotes a system with a reciprocal interaction of forces. X_1 affects X_2, which in turn affects X_1; once we compare the range of change, we can find a stable equilibrium—or not, in which case it would be an unstable one.

In continuing with this line of reasoning, and building upon this found range of variability, Ashby stated that all systems in equilibrium have a neutral point. This is pointed out as an important claim, since the stable equilibrium of a system would depend on the amount of variation away from this neutral point: if the disturbance pushes the system out of the said neutral threshold, then it becomes unstable. In fact, this ontological leash, within which the system can still be said to be in stable equilibrium, is given by Ashby the name of "range of stability".[16]

---

[14] "Every body persists in its state of being at rest or of moving uniformly straight forward, except insofar as it is compelled to change its state by force impressed" (Newton 1687, p. 416).

[15] Ashby 1940, p. 479.

[16] Ibid., p. 481.

Since the whole point of the argument is to adequately articulate the notion of a machine in order to defend its claimed isomorphism with the nervous system (or any other living organism), Ashby readily acknowledged that one likely criticism would be that such circuitry is way too simple for it to have any hope of being compared to a living entity. To this, he advanced the response that such "simple" circuits can very quickly become complex—and in an extreme manner. Consider the following case:

$$x_1 \rightleftarrows x_2 \rightleftarrows x_3$$

Or even...

One could readily notice that the modification of just one variable will alter the whole system. And so, if we have more variables, and each one of them is itself modified, the compound circuitry can quickly engage in dramatic complexity. The adaptive circuitry of a natural machine (a living organism), thus, should no longer be regarded as belonging to a very high complexity of a different *kind*—it is rather different in a matter of *degree*. Ashby was reassured that what he found has serious implications for our properly mechanistic understanding of life. He exhorts us to see that...

> there is a point of fundamental importance which must be grasped. It is that stable equilibrium is necessary for *existence*, and that systems in unstable equilibrium inevitably destroy themselves. Consequently, if we find that a system *persists*, in spite of the usual small disturbances which affect every physical body, then we may draw the conclusion with absolute certainty that the system must be in stable equilibrium. This may sound dogmatic, but I can see no escape from this deduction.[17]

---

[17] Ibid., p. 482.

A system's equilibrium has to remain within the neutral point for it to remain stable, so that the "range of variability" occurring in its circuitry would allow for its proper adaptation to the changing environment. Ashby is well aware that when such process occurs in the natural world, it is called by a familiar name: *survival*. The mechanistic bridge with living organisms has now been established. The most important feature exposed by living entities, the very feature that accounts for its survival, would seem to be a characteristic of matter itself—as long as it is a coping system that remains stable. In Ashby's own words...

> The moment we see that "adaptiveness" implies a circuit and that a circuit implies an equilibrium, we can see at once that this equilibrium must be of the stable type, for any unstable variable destroys itself. And it is precisely the main feature of adaptive behaviour that it enables the animal to continue to exist... Vast numbers of variables associated with the animal are all in stable equilibrium. Not only is this so as an observed fact, but it is clear that it must be so because any variable or system in unstable equilibrium inevitably destroys itself.[18]

Having realized this organism-machine identity in terms of structure, Ashby went on to deal more specifically with the characteristics of its coping behavior. In a subsequent article five years later,[19] he hypothesized about the possible processes that might take place when an organism copes with an altering environment in order to survive. He proposed that a living entity engages in a trial-and-error operation, altering variables in its inner circuit until it finally finds the right neutral "sweet spot", where it can remain stable—*alive*—exactly as when a coping machine reaches equilibrium.

The trial-and-error coping in the animal portrays the following characteristics: 1) It is called upon only when the environment provides a hostile milieu for the organism; 2) Each tried option remains for a determinate period of time, until it is no longer satisfactory; 3) When the hostile environment remains, the trial-and-error dynamic does not stop; 4) The next attempted option is randomly chosen, not necessarily being a better one; 5) When a tried choice provides an adequate coping,

---

[18] Ibid., p. 483.
[19] Ashby 1945.

then the system stops "hunting" and remains stable. This would entail an animal successfully surviving.[20]

Ashby is here introducing an astonishing suggestion. This trial-and-error procedure, he proposes, is not unique to living entities; quite on the contrary, "it is an elemental and fundamental property of all matter".[21] In fact, this biological operation finds its mechanical counterpart in the phenomenon of a machine "breaking". To push this point through, Ashby deemed it necessary to squarely tackle the notion of a machine, carefully picking it apart. This might well be the first time in history that the concept of a machine *per se* was directly addressed.[22]

Ashby indeed had at this stage an incipient definition of a machine— one that would be increasingly fine-tuned throughout the subsequent years. A machine, he claims, is "a collection of parts which (a) alter in time, and (b) which interact on one another in some determinate and known manner".[23] We are in front of characteristically machine-like features of change, reciprocal interaction, determinism, and measurability. The parts can be called variables, since they are measurable. The alteration of these variables in themselves does not necessarily modify the overall constant of the system in equilibrium; however, a change in

---

[20] Quoting the American geneticist Herbert Spencer Jennings (1868–1947), Ashby subscribes to the idea that

> Organisms do those things that advance their welfare. If the environment changes, the organism changes to meet the new conditions... If the mammal is cooled from without, it heats from within, maintaining the temperature that is to its advantage... In innumerable details it does those things that are good for it (Jennings 1906, p. 338).

Ashby would also acknowledge thinking in the same vein as the Scottish physiologist John Scott Haldane (1860–1936), who recognized homeostatic endeavors of organisms in his studies on respiration: "Biology must take as its fundamental working hypothesis the assumption that the organic identity of a living organism actively maintains itself in the midst of changing external circumstances" (Haldane 1922, p. 391).

[21] Ashby 1945, p. 13.

[22] See note above in this section on Hertz.

[23] Ashby 1945, p. 14.

the organization of the network of parts does. Thus, the identity of a machine has to do less with changes in its parts, and more with the change in the constants of the overall array of variables—in other words, with its internal *organization*.

When a machine "breaks", Ashby reminds us, the entity is still a machine—but another type of machine, since its internal organization has changed. A "break" is thus a change in the constant of the overall network of variables that conjointly forms the machine. And this break has a reason to exist. When a coping machine, due to an external disturbance, is pushed out of its neutral point of stability (equilibrium), its variables will individually change so that in group they compensate for the attempted alteration of the network. This compensation brings back the machine to its stable equilibrium, effectively steering it away from the overall effect of the disturbance:

> A machine which has available an indefinitely large number of breaks depending on configurations closely similar to one another will inevitably change its internal organization spontaneously until it arrives at an organization which has an equilibrium with the special property that it avoids those configurations.[24]

When the disturbing force is stronger than what the variables are able to compensate—and thus collectively push back—then a change in the overall network occurs. The constant behind the equilibrium point then changes to *another constant*—constituted by the overall change of altered parts—so that the machine can persist in stable equilibrium. Each break will send the array of parts hunting for a newly organized internal structure, until another constant is reached and a stable equilibrium is again found. If another constant is not eventually reached in a finite timeline, the machine does not break: it gets destroyed. It does not survive.

Now we can better see the ontological bridge with a living organism firmly set in place. If we recall, an animal has to re-wire its internal circuitry in order to cope with an altering environment. This is done by means of a trial-and-error procedure, its variables individually changing until the overall network of parts finds a stable equilibrium with the modified environment. This constitutes the characteristic animal adaptiveness.

---

[24] Lack of "commas" in the original. Ashby did not seem to be particularly fond of generous punctuation.

The attained equilibrium will persist until another change will make the organism "break", once again until a new successfully coping self-organization is achieved, and so on. This is survival. And this is the manner in which a successfully coping machine persists. Organism and machine, in virtue of both having one quintessentially defining behavior in common— self-organization in order to survive—can be said to be, effectively, one and the same.

THE HOMEOSTAT: A LIVING MACHINE

In an article published two years later,[25] another previously suggested point of no small importance for Ashby's subsequent proposal is reinforced. A machine's ontological counterpart in nature is, according to him, a living organism *and* its environment—the animal alone is an incomplete machine. *Both* organism and environment, engaged in dynamical interaction, constitute a machine. In fact, that will be a key distinction for Ashby's next conjecture regarding the fact that a certain type of living organism (the nervous system) successfully adapts to change— rather than being caught up in an endless internal chaotic loop, entailing what certainly occurs at times: death.

Having already established the isomorphism between working machine and living being, Ashby did not waste time in expressing a keen interest in having a particular type of organic entity understood as a machine: the nervous system (and its environment). In it, he advances, "the 'variables' are mostly impulse-frequencies at various points in the nerve-network. And as the impulses at one point affect in various ways the impulse-frequencies at other points, the nervous system is a dynamic system par excellence".[26] Citing empirical studies of his time that in his mind were suggestively friendly to his view, Ashby felt that he succeeded in providing a full machine-based explanation for how (at least basic) living entities survive.

Apparently not entirely content with what he had so far accomplished theoretically, in elegantly cybernetic fashion Ashby had only one year later (1949a) publicly acknowledged having constructed a

---

[25] Ashby 1947a, pp. 56–57.
[26] Ashby 1947a, p. 57.

machine that would instantiate his hypothesis "in the flesh".[27] He built what he baptized as the Homeostat—with obvious reference to his intellectual mentor Walter Canon's studies in biological homeostasis. This machine, which was to become a celebrity among hardcore cyberneticians,[28] was built from WW2 spare military parts.

Echoing a salient characteristic of British cybernetic pursuits, Ashby's machine was gradually built during his free time, partly at home and partly at the storage area of the hospital where he worked. There was no formal institutional funding behind its construction.[29] The apparatus consisted of four Royal Air Force bomb control switch gear kits[30] mounted on four aluminum boxes, each of which connected to a pivoting needle with a concave trough filed with conducting liquid (water). Each control gear switch, or uniselector, was previously hardwired with random values,[31] constituting the particular threshold that the current had to reach—or not—in order to be let through. The four units were electrically interconnected among themselves. Each had 25 possible combinations. When the current of electricity passed through them, the overall result was that each of the uniselectors' floating needles would reach a middle point in the trough, and remain

[27] "In the metal", to be precise: Ashby started constructing it in 1946 and presented the machine at a meeting of the Electroencephalographic Society of Britain in 1948 (Ashby 1949a).

[28] As opposed to the famed tortoises build by Grey Walter, which enjoyed widespread fame among the general public (Boden 2006, pp. 228–230).

[29] Eventually Ashby was given a professorship at the University of Illinois at Urbana-Campaign, but at this time it is safe to say that "classical" cybernetics was already spiraling into oblivion. However, funding for his second machine (Section "Un-cybernetic DAMS: From Behavior into Structure", below) was secured (Pickering 2010, p. 310).

[30] Ashby 1928–1972, p. 2432. These are stepping switches or uniselectors— electromechanical devices that allow for the input and output of electric current via a chosen built-in channel at a time.

[31] These were taken straight from a table of random numbers that Ashby found in Fisher and Yates's *Statistical Tables*, originally published in 1938 (Fisher and Yates 1963, pp. 134–139).

there. The four interconnected uniselectors formed the whole homeostat.[32]

The peculiarity of the machine relied on the following feature. If one were to modify the connections, say "re-wire" the machine in another way—inverting the current, connecting the uniselectors in different order, messing with the needles, turning off one of the uniselectors, etc.—the needles would appear to go hunting for a different position, trying out different states, until they would all settle back into each one of their middle points in their troughs. Ashby's colleagues would find themselves frantically re-wiring the machine several times, in different ways, just to witness somewhat in awe how the machine would "re-organize" itself, gradually returning to its initial point of stable equilibrium at the center—

---

[32] Ashby describes the construction of his machine in the second—substantially revised—edition of his book *Design for a Brain* as follows:

> The Homeostat... consists of four units, each of which carries on top a pivoted magnet... The angular deviations of the four magnets from the central positions provide the four main variables. Its construction will be described in stages. Each unit emits a D.C. output proportional to the deviation of its magnet from the central position. The output is controlled in the following way. In front of each magnet is a trough of water; electrodes at each end provide a potential gradient. The magnet carries a wire which dips into the water, picks up a potential depending on the position of the magnet, and sends it to the grid of the triode... If the grid-potential allows just this current to pass through the valve, then no current will flow through the output. But if the valve passes more, or less, current than this, the output circuit will carry the difference in one direction or the other...
>
> Next, the units are joined together so that each sends its output to the other three; and thereby each receives an input from each of the other three... But before each input current reaches its coil, it passes through a commutator..., which determines the polarity of entry to the coil, and through a potentiometer..., which determines what fraction of the input shall reach the coil. As soon as the system is switched on, the magnets are moved by the currents from the other units, but these movements change the currents, which modify the movements, and so on (Ashby 1960, pp. 100–102).

equidistant from either extreme, 45 degrees at each side. Ashby would later succinctly describe its behavior like this:

> On top of each box is magnet [*sic*] which can turn on a pivot. The central position counts arbitrarily as "optimum"; at 45° on either side is the "lethal" state the brain must avoid, and it is at that position that the relay closes to make the uniselector change to a new position. If it is stable, the needles come to the center; if they are displaced, they will all fluctuate but they will come back to the center. That behavior corresponds to the behavior of the adapted organism.[33]

As we can see, Ashby expectably regarded this behavior as being one and the same with that of a living entity *surviving*. Following his previous understanding of a dynamical machine as conformed by both animal *and* environment, part of the homeostat's configuration would be the organism per se, and the other would be its threatening milieu.[34] In fact, the titles of the first mainstream publications that featured the machine were "The Electronic Brain"[35] and the "The Thinking Machine"[36]—in *Radio Electronics*, a contemporary popular electronics magazine and the famed *Time*, respectively.

Three years later, Ashby's book *Design for a Brain* (first edition 1952a) gave a preeminent role to his machine, locating it almost at the axis of how and why we should see the machine-based hypothesis as very likely true: it is presented as an instantiation of a theorem. That same year, equipped with his "cybernetic monster",[37] he was finally invited by the "cybernetic group" to the Macy conferences in New York. This occasion will turn out to be the first and last time that he had the opportunity to attend the cybernetic Mecca of intellectual exchange. As mentioned in Chapter 3,

---

[33] Ashby 1953, p. 97.

[34] What part of the machine would be the animal and what part would be the environment was entirely up to the operator. We could regard, say, one uniselector as a small animal and the remaining three as a big environment, or vice versa. Indeed, Ashby referred to the Homeostat as "a machine within a machine" (ibid, p. 96).

[35] Ashby 1949b.

[36] Ashby 1949c.

[37] Pickering 2005.

of the ten conferences that took place in a span of less than a decade (1946–1953), only the last five enjoyed published proceedings. Since this was the penultimate one, we have a fairly detailed account of what happened on the day of his presentation.[38] It is helpful to complement it with Ashby's personal diary entries,[39] in an effort to shed light on what took place that day. What happened in there represents a somewhat strange and unique occurrence in what regards twentieth-century science gatherings.

THE MACY PRESENTATION: KEEPING CYBERNETICS HONEST

Right from the opening remarks of his talk, Ashby did not hide both the bold epistemic status of his proposal—namely, understanding the essence of life as identical with machine behavior—and the audaciously attributed ontological status of his machine—as inhabiting a middle-ground between the living and the non-living...

> I must say unambiguously what I mean by "alive". I assume that if the organism is to stay alive, a comparatively small number of essential variables must be kept between physiologic limits. Each of these variables can be represented by a pointer in a dial... This, of course, is in no way peculiar to living organisms. An engineer, seated at the control panel in a ship, has exactly the same task to do. He has a row of dials some of which represent essential variables in the ship, and it is his business to run things so that the needles always stay within their proper limits. The problem, then, is uniform between the inanimate and the animate.[40]

Ashby went to great lengths in trying to explain what he was about to show. He first wanted to set the proper context so that when the homeostat was introduced, the participants were already familiar with the issue to be addressed—namely "how the organism manages to establish homeostasis in the larger sense".[41] This was not an easy feat. Ashby had to carefully convey, in a very succinct way, roughly the gist of the research

---

[38] The meeting took place on Thursday, March 20th, 1952, and it was subsequently published as Ashby 1953.

[39] Ashby 1928–1972, pp. 3732–3736, 3738–3744.

[40] Ashby 1953, p. 73.

[41] Ibid.

mentioned at the beginning of this chapter. An important aspect of this aim was to establish the feasibility of an unambiguous setting of the problem—task which some participants clearly saw and agreed upon, while others did not quite get.[42]

Ashby started asking whether or not an animal's changing environment can be specified in a clear and rigorous way. The emphasis would not be placed on the environment's complexity, but on its degree of determinateness. This would allow him to on with his proposal. But for this to happen, Ashby had to first agree on that point with the audience. And despite the common cybernetic credo of the conferees, it encountered some resistance at first,[43] probably due, to some extent at least, to the difficulty in understanding where Ashby was actually going with all this. Eventually they settled in agreeing that a sharp and clear cut structuring of the problem was indeed possible ...

> *Ashby:* Let the environment be represented by the operator, E. The organism's problem is to convert the brain into an operator, which might be represented by E^{-1} ... The formulation with E and E^{-1} is merely meant to state the problem in a form that is clear physically and

---

[42] Let us recall that the Macy conferences were radically interdisciplinary, its participants coming from several distinct fields encompassing mathematics, engineering, biology, psychology, anthropology and others. Thus although all were sharing a cybernetic interest, they were not all comfortable with Ashby's terminology. Throughout the presentation, one could notice some participants struggling with getting the conceptualization of the homeostatic problem right. Ashby was merely preparing the context for the introduction of the machine, and complications seemed already happening (See Ashby 1953, pp. 73–75).

[43] Although eventually the majority of the audience agreed with Ashby on the ontological feasibility of accepting the variables E and E-1, one could perceive that the trained engineer Julian Bigelow was going to be Ashby's nemesis throughout his talk. Regarding Ashby's plea to accept the ontological feasibility of the variables E and E-1 in order to continue with the presentation of his proposal, Bigelow had this to say:

> *Ashby:* The chief thing that I'm interested in is the question, if this system is grossly non-linear and if there are delays ... can this operator E be given a meaningful and rigorous from?
> *Bigelow:* In general, no (Ashby 1953, p. 75).

mechanistically . . . E^{-1} is fairly straightforward, if E^{-1} does not exist, the organism cannot find a solution and dies. If it does exist, then the animal may perhaps find several solutions. But what I want to know is whether E can be given a fully rigorous definition.[44]

Wiesner: [45] That is an operation we would agree it can be done.

Pitts: It is certain that it could be.

Ashby: You think it could be?

Wiesner: By definition, yes.[46]

Subsequently Ashby shared the reason why focusing right from the start on the adequate addressing of the environment was a mandatory requirement. Without rigorously treating, articulating and then factoring in the environment, little would afterwards be intelligible when talking about the brain. Thus, a deep study of the organ ought to have an equally profound study of its corresponding containing milieu. He claimed that

> there can't be a proper theory of the brain until there is a proper theory of the environment as well . . . Without an environment, "right" and "wrong" have no meaning. It is not until you say, "Let's give it E_{47}, let's see what it will do against that", it is not until you join the brain on to a distinctly formulated environment that you begin to get a clear statement of what will happen.[47]

Ashby's position regarding the nature of such environment would appear quite reasonable to some, while utterly unrealistic to others—depending on one's philosophical leniencies. The animal environment's elements

---

[44] One should guard oneself from being misled into thinking that Ashby means "mechanistic" in a solely heuristic way. In accordance to the rest of his philosophy, when an *explanans* was rigorously mechanical, the *explanandum* happened to be a machine. The behavior of the entity, and not its structure, determines whether or not that thing is a machine. That is part of the crux of his contribution. And that is also the source of the problems to be addressed below.

[45] Jerome Wiesner (1915–1994) received his Ph.D. in Electrical Engineering from the University of Michigan and went to work at MIT's previously mentioned Rad Lab. He was a science consultant for the Kennedy administration. Later in his life, much like Norbert Wiener, he became a vociferous advocate against war.

[46] Ashby 1953, pp. 75–76.

[47] Ibid., p. 86.

were found by Ashby to be existent, at least temporarily, in a physically discrete and non-continuous manner. However, the Humean chunks he was dividing reality into were not to be understood as effectively isolated from each other; rather, they were to be regarded as *not totally and continually* interconnected among themselves...

> There is a form intermediate between the environment in which everything upsets everything else and the environment which is cut into parts. This intermediate type of environment is common and is of real significance here; it is an environment that consists of parts that are temporarily separable, and yet by no means permanently separable...[48]

This created some tension in the audience. Bigelow replied that "[o]ur environment doesn't consist of that sort of phenomena in very many ways. The statement you are making amounts to something that mathematically sounds independent but does not exist in the real world".[49] It may have been the case that Ashby was not being properly understood, since he did make the remark (above) that environmental factors, even if we can dissect them to some extent, are "yet by no means permanently separable".[50] And this misunderstanding may had not been exclusive to Bigelow. For instance, Heinrich Klüver[51] politely interrupted to say:

> Your scheme should be of particular interest to psychologists, because unlike other models we have encountered here, it stresses environmental factors and the interrelations between animal and environment. I did not quite follow what was implied in your statement that constancies may function as barriers.[52]

As addressed at the beginning of this chapter, once the coping machine, in trying to accommodate itself to the changing environment, reaches an over-all stable equilibrium, this attained state remains constant, until another change is such that the parts/variables are obliged to go hunting again—in

---

[48] Ibid., p. 82.

[49] Ibid., p. 84.

[50] Ibid., p. 82 (quoted above).

[51] Heinrich Klüver (1897–1979), German-American psychologist who received his Ph.D. from Stanford University.

[52] Ibid., p. 86.

which case we can say that the machine goes through a "break". But until then, this threshold, this "ontological leash" that such parts enjoy, allow them to tolerate and push back the changes, until the overall pressure is such that a break has to occur—so that the animal can again re-wire itself. Also, having in consideration that in order to have a flow of information, "change" must be present, a variable locked into an achieved constancy will effectively block the flow of information.[53] Accordingly, and putting emphasis on the hunting parts, each trying to be adequate to the changing circumstances, Ashby patiently answered that

> If a system is composed of variables that often get locked constant, it tends to cut the system into functionally independent sub-systems which can join and separate. Instead of being a totally interlaced system with everything acting on everything else, it allows sub-systems to have temporary independencies.[54]

Hopeful to have cleared out the point, Ashby went on with his presentation, just to be interrupted with clarifying questions by other participants every so often—which he patiently addressed. The most engaged into the discussion were three authors already familiar to us (McCulloch, Pitts, and Bigelow), and Wiesner. Ashby, probably acutely self-aware of the next, bolder step he was about to take, did not hesitate in answering each and every question—sometimes briefly, other times lengthily, but always firmly.

What could be deemed in contemporary academic ambiences as a hostile milieu for debating was seemingly just part of a typical heat of the discussion in the 1950s.[55] In fact, not all remarks showed puzzlement, in the spirit of "what is it that he is talking about". Some were enthusiastic about the wild cues that were seemingly being suggested. Gerhardt von Bonin, [56] for

---

[53] As seen in Chapter 5, Section "Machines can be Teleological", Claude Shannon's foundational paper (Shannon 1948), despite the secrecy that surrounded it when it was first written, was already very well known in the cybernetic circle.

[54] Ashby 1953, p. 86.

[55] Ashby wrote down of this day in his diary as being "highly successful" (Ashby 1928–1972, p. 3732).

[56] Gerhardt von Bonin (1890–1979) was a German-American neuroscientist, professor at the University of Illinois.

example, in referring to Klüver's above participation and Ashby's response to it, added sympathetically . . .

Von Bonin: To pick up the remark Klüver made a moment ago, the animal appears to break down the environment into certain patterns which seem to develop in his mind. Something goes on in its brain and then that structures the environment. It perceives the environment in a certain pattern, set by its brain, so that it can deal with it.

Ashby: I assume that the brain works entirely for its own end.

Von Bonin: Oh surely.

THE BACKLASH: UNFORESEEN CONSEQUENCES OF A BEHAVIOR-BASED ONTOLOGY

Some of the members attending the talk were taken aback by the claims of the homeostat behaving isomorphically to an animal in reciprocal interaction with its environment. It was not only that unlike Grey Walter's tortoises and Claude Shannon's mice this machine did not formally resemble any actual animal. What worried them was what was being suggested as entailing from the machine's random looking for a stable equilibrium after its uniselectors were tampered with. Even if they would be willing to accept that some primitive organisms do indeed resort to random trials in order to cope with a changing milieu, it seemed that the whole image of four interlocking uniselectors mounted on a table and allegedly behaving animal-like, was more than what they were willing to accept. Bigelow, somewhat in confusion, expressed that the homeostat "may be a beautiful replica of something, but heaven only knows what".[57] Others sided with him . . .

Hutchinson: [58] Do you interpret the ordinary sort of random searching movement of one of the lower animals as something that is comparable to the uniselector?

Ashby: Yes.

---

[57] Ashby 1953, p.95.

[58] George Evelyn Hutchinson (1903–1991) was a British zoologist and ecologist who spent most of his professional life teaching at Yale University—although he never earned a Ph.D.

> *Hutchinson:*   It doesn't seem to me that the setup gives any particular clear suggestion that this is comparable to the way invertebrates do behave.
> *Bigelow:*   Agreed.[59]

It would seem that some metaphysical attachment to the nature of living beings was still not being let go. And this situation was happening among those very individuals whose aim, as proper cyberneticians that they were, was to dismantle any me taphysical remnants behind a vitalistic impetus that might differentiate in kind—and not degree—a living organism from a machine. Ashby seemed to have noticed it. Without a sign of dubitation, he was firm in eroding bit by bit the seemingly still entrenched epistemic attitudes being witnessed at work. Faithful to "cybernetically correct" belief, Ashby did not lose focus and rehashed on *that* aspect of both machine and organism that is core of cybernetics parlance: *behavior*. For all ends and purposes, an organism behaves in the way the homeostat behaves—literally.

Beginning from Bigelow's (and company) 1943 foundational article,[60] that pillar of cybernetics was never shaken. Ashby merely rehearsed what they already knew, and indeed, defended[61]—albeit probably unaware of the ultimate consequences that would naturally emerge from such knowledge.[62] Randomness in animal adaptation was probably

---

[59] Ibid., p. 98.

[60] Rosenblueth et al. 1943.

[61] Dupuy 2000, pp. 148–155.

[62] As seen above in this chapter, it would seem that Ashby had faithfully adhered to cybernetic ideas in a manner that preceded Wiener himself, judging from his 1940 article (Ashby 1940). However, importantly for this work, Ashby does acknowledge that a profound, fundamental insight into the nature of a machine is first given by Wiener:

> I see that my idea that maths is a dynamic process, and that the Russellian and other paradoxes are simply the machine going to a cycle, has already been stated by Wiener. Maybe I got the idea from him.

Ashby goes on to quote a line from Wiener's *Cybernetics*: "... the study of logic must reduce to the study of the logical machine ... " (Ashby 1928–1972, p. 6023).

the characteristic that helped the most in establishing the ontological link between machine and living being. That was the way in which a living organ, as adaptationally sophisticated as a brain, could be said to behave in a machine-like manner. Machines can certainly adapt: the name for that in the physical world is equilibrium, as Ashby was careful to establish in his first published cybernetic article addressed earlier.[63] And so, he pushed the idea of randomness as properly animal, and properly homeostatic, until the end:

Hutchinson:	I am not altogether convinced that that searching [an ant looking for sugar cubes dropped on a table] is in any way comparable to what we have seen here.
Ashby:	But don't you agree that if you disturb a living organism with some vital threat and the organism takes steps to bring itself back to normal physiological conditions, that is perfectly typical of vital activity?
Hutchinson:	Yes, but I think when it does that, there is not the fluttering oscillation that you first described. Isn't that inherent in the system?
Ashby:	Well, some sort of random trial is necessary. Surely, random trying in difficult situations has been described over and over again from Paramecium upwards.
(...) Quastler: [64]	I believe Dr. Ashby would not claim that random searching was the only mechanism by which an organism finds a normal or favorable condition.
Ashby:	I don't know of any other way an organism can do it.[65]

A further point of contention derived from the above remarks was the issue of learning as an outcome of a random process. They seemed to be uncomfortable with the idea that a change of behavior—a resulting accommodation—coming out of a random search could count as learning proper. Once again, it seemed that the very proponents of the treatment of machines and animals

---

[63] Ashby 1940.

[64] Henry Quastler (1908–1963) was an Austrian-American physician and radiologist at the University of Illinois. After WW2 he did pioneering research on the effect of radiation in living tissue.

[65] Ashby 1953, p. 99.

under the same physical laws could not foresee the metaphysical backlash that could follow. In cybernetics, let us recall, behavior is the sole and privileged communicator of any change of a state in a given living entity. In Bigelow's (and company) famed 1943 article, one of its principal goals is indicated as attempting to "define the behavioristic study of natural events and to classify behavior".[66] This most important notion is subsequently referred to as "any change of an entity with respect to its surroundings".[67] Further clarifying, it is claimed that "any modification of an object, detectable externally, may be denoted as behavior",[68] enriching the otherwise reductive understanding of behavior coming from psychology.[69] A perfectly feasible outcome from this established framework is accepting as meaningful the variation from state A to state B in a given organism after it deals with a change of its environment. It did not take much for associating the semantics of a learning process—itself already understood behaviorally, independently of cybernetics—to the meaningful change of state of a machine. Change of behavior after coping with a modifying milieu *is* learning. And this necessary development is what makes the following lines somewhat striking in their naïveté, and fascinating when confronted with Ashby's calm, perfectly cybernetic-abiding answers:

> *Bigelow*: Sir, in what way do you think of the random discovery of an equilibrium by this machine as comparable to a learning process?
>
> *Ashby*: I don't think it matters. Your opinion is as good as mine.
>
> *Bigelow*: But I'm asking, if you place an animal in a maze, he does something which we call learning. Now, in what way does this machine do something like that?

---

[66] Rosenblueth et al. 1943, p. 18.

[67] Ibid.

[68] More specifically, it is stated:

> Given any object, relatively abstracted from its surroundings for study, the behavioristic approach consists in the examination of the output of the object and of their relations of this output to the input... The above statement of what is meant by the behavioristic method of study omits the specific structure and the intrinsic organization of the object (ibid.).

[69] See Chapter 2, Section "The AA-Predictor".

(...) Wiesner: I think that [the mechanical mouse of] Shannon and Marceu came close to having a learning machine.

Bigelow: That is the point. This machine finds a solution, I grant you—and please call it anything you want; I'm not trying to criticize your language. I merely wonder why finding a solution necessarily implies that it learns anything, in your opinion.

Ashby: I think that the word learning, as we understand it in the objective sense, without considering anything obtained introspectively, is based on observations of this sort of thing happening.[70]

Ashby seemed perfectly aware that the pillars upon which he was relying were as strong as they could get, in terms of being flawlessly consistent with essential cybernetic tenets. It is as if he was clarifying to the cyberneticists themselves what their own beliefs entail—besides letting them know at the same time how perfectly comfortable he is with those underlying frameworks. Further, Ashby seems willing to champion these tenets and apply them all the way down, regardless of the consequences. He does not look in the least afraid of forever losing a metaphysical anchor that might had been in some non-scientific way beneficial for the sub-group of machinery of which he is himself part. Faced with the awkward complaints being generated by the cybernetic group itself, Ashby cut clean through their malaise, and confronted them with the expectable effect of their own machine-based ontology and behavior-based epistemology...

Young: [71] [The homeostat] doesn't show change of state. The essence of learning is that the system that has been through a procedure has different properties than those it had before.

Ashby: Would you agree that after an animal has learned something, it behaves differently?

Bigelow: Yes.

Ashby: Well, the homeostat behaves differently.[72]

---

[70] Ashby 1953, p. 103.

[71] John Zachary Young (1907–1997) was a British zoologist and neurophysiologist, who applied research in animal nervous systems to injured humans after WW2.

[72] Ashby 1953, p. 103. Bigelow was, however, still reticent right until the quoted above behavioristic gauntlet was thrown. He told Ashby:

This seemed to do it. The change in attitude was almost immediate. Despite the painful remarks thrown around that marred almost the entirety of Ashby's talk, agreement became suddenly consensual, to the point that the audience, almost leaving Ashby aside, took the role of vindicating his proposal. As heated as debates can come, the attendees took up the torch and continued engaged in conversation among themselves, not even allowing the speaker to formally end the presentation.[73] All things considered, and checked out with the core of cybernetic mandates, they had to agree with the claimed facts that 1) the behavior of the homeostat is identical to that of an organism coping with its changing environment and 2) the homeostat, for all intents and purposes (literally), learns. And so they became an audience geared into acceptance mode:

Bigelow:	...Do you consider this a learning device as it stands?
Pitts:	Yes, I should say so.
McCulloch:	Would you consider Shannon's mechanical mouse a learning device?
Bigelow:	Surely.
McCulloch:	Then this is a learning device in just that sense Shannon's mouse learns how to run *one* maze.[74]
(...) Bateson:	I should like to put a question to our ecologist: if an environment consists largely of organisms, or importantly of organisms, is not the learning characteristic of Ashby's machine approximately the same sort of learning as that which is shown by the ecological system?
Hutchinson:	Yes, definitely it is.[75]

...if you have a box with a hole in the bottom and a ball bearing inside and keep banging at it, the bearing will fall through. That is in your sense, a way of learning (1953, p. 104).

To this, Pitts interjected:

The homeostat is a bit better than that (ibid.)

[73] As per the meeting transactions (Ashby 1953, p. 108).

[74] Ashby 1953, p. 105. Italics added.
     The reason why "one" is important in this context will become clear below.

[75] Ibid., p. 106.

These were the last interventions. Ashby wrote in his diary that the presentation was "highly successful"[76]—although acknowledging the distress of some.[77] From a historical perspective, it could be said that this conflicted exchange of ideas set up the environment for what could *effectively* count as the last meeting for the cybernetics group. The next meeting lacked so many key members and was so scattered and out of focus, that the editor of the proceedings refused to publish them—eventually, they were indeed published but only the papers, without the discussions.[78] One cannot help but wonder if that meeting was (perhaps more than) a symbolic *cout d'état* for a body of ideas that was already receiving a battery of attacks from different flanks.[79] What happened that day might have been just the tip of the iceberg regarding the deeply problematic ontology on which cybernetics was resting. That ontology was already leaking concerns at various levels of the cybernetic enterprise—however they were not identified as a whole, and anyhow there never was a unified effort to patch them.

Four years later Ashby would give a more detailed account of the mechanism behind the homeostat, paying special attention to the problematic feature of its random search. As footnoted above, this feature used a published table of random numbers by Fisher and Yates. The fact that the picked numbers from the table were fixed in the machine, occasioned some trouble in the audience, in regards to the veracity of the claim that the machine was searching randomly. However, this fixation

---

[76] Note above (Ashby 1928–1972, p. 3732). Ashby seemed to be particularly delighted with Walter Pitts' siding with him in regards to the proper framing of the homeostatic problem specified above that his machine seemed to portray: "Walter Pitts assured me that E is certainly well defined if it is assumed to be some neutral physical mechanism, for the latter's factuality ensures proper definition" (ibid., p. 3738).

[77] For instance, he says this of Julian Bigelow's intervention: "At the Macy Conference, Bigelow was very insistent that there was no virtue in random searching as against systematic searching" (ibid., p. 3744).

[78] See Chapter 2, Section "The Macy Conferences", toward the end.

[79] As one could induce from the conflictual nature of the Macy debates (Chapter 2, Section "The Macy Conferences", regarding the 7th Macy Conference). Admittedly, it could also be seen as the inherently difficult nature of the interdisciplinary trading of ideas.

was also an essential part of the necessarily deterministic nature of the machine, as Ashby indicated at some point.[80] But it was clear that the idea of order achieved out of chaos was not sitting well among the cyberneticians present at that gathering.[81]

UN-CYBERNETIC DAMS: FROM BEHAVIOR INTO STRUCTURE

Another issue that Ashby would go on to treat was related to the criticism of the homeostat being effectively too primitive to mimic higher vertebrates' behavior—specifically, mammals. One critique that elicited his agreement was the claim that the homeostat probably was indeed capable of autonomously reaching self-equilibrium due to the fact that, after all, 4 uniselectors combinable in 25 different ways among themselves (and accounting for a bit more than 390,000 possible combinations), would still fall very short of representing the extreme adaptive richness of higher animals. Ashby acknowledged the input and granted this point. In fact, he already was considering constructing something capable of more complexity than the homeostat—with more possible internal combinations than the homeostat's less than 400,000—although operating under the same principles. Ashby was aware that he needed to construct a machine with more complexity: one that, if possible, should show a behavior that would genuinely surprise its designer—show something that was not put in there by the inventor. Indeed, almost two years before the presentation, he already began thinking about constructing a more complicated machine.

---

[80] Jerome Wiesner took the "random search" for stability performed in the homeostat as grounds for stating that the machine was "randomly operated"—which it was not. There were clear thresholds in each stepping switch, determined by hardwired random numbers that had to be reached, in order for the current to go through to the next uniselector, until they would all return to a stable state of equilibrium. Thus, Ashby interjected: "Just a minute. I don't like the term 'randomly operated.' The machine's behavior depends, determinately, on whether its essential variables are inside or outside their proper limits" (Ashby 1953, p. 96).

[81] However, this epistemic position might had already crawled within cybernetics by that time, as it is noted by Ashby's total comfort with the idea. See Chapter 5, Section "Machines Can Be Teleological" (toward the end).

I have to note a critical day. Ever since the homeostat was built (May '47) I have been thinking about the next machine, trying to find a suitable form for a more advanced machine.[82]

Its longer name was going to be *Dispersive And Multi-stable System* (DAMS). The report on the construction of such a machine appeared at the footnote of the first edition of his *Design for a Brain*—and disappeared in later editions.[83] Not much is known about this machine, other than the entries recorded in his diary.[84] The idea behind giving DAMS a more complex nature relied on the possibility of letting the machine isolate some of its parts when they would have reached adaptation, and letting the rest of the machine keep hunting for other "friendly" variables. What was to be gained from this is that it would no longer be necessary to re-wire the whole machine to prepare it for the next task. As seen in the previous section, given the benefit of the doubt, Ashby was granted that the homeostat portrayed the same level of learning behavior of the cybernetic mice, which could sort out a maze *once*, before getting ready to get re-wired for another task.[85] The neon tubes installed in DAMS allowed electric current to go through only when they would reach a certain threshold, remaining inert and nonconductive when the threshold was not met. This was an improvement over the homeostat's wires—which were always conductive—and affected the whole system. Always. Truer learning could allegedly be achieved when already reached adaptations were kept and other portions of the organism continued coping. However, finding the right form of design for this more

---

[82] Ashby 1928–1972, p. 2950.
[83] Pickering 2010, pp. 122–123.
[84] Ashby 1928–1972, Other Index, 4: DAMS.
[85] The excerpt of the relevant dialogue portrayed above reads thus:

> *McCulloch*:   Would you consider Shannon's mechanical mouse a learning device?
> *Bigelow*:   Surely.
> *McCulloch*:   Then this is a learning device in just that sense Shannon's mouse learns how to run *one* maze (1953, p. 105. Italics added).

(The reason why I previously italicized the word "one" should now become apparent)

sophisticated device was not an easy feat for Ashby—in fact, it eventually proved to be more than what he could handle.[86] At a certain point, believing that being stuck in the "design stage" of the construction ironically betrays the whole purpose of a self-accommodating ultra-stable system, he foregoes the design aspect, puts the pieces together, and waits to see what happens . . .

> Today it occurred to me that this is not the proper way. According to my own "theory" all this design is irrelevant; what I should do is to throw the design to the winds, join all at random, and trust. I should not attempt to follow the consequences of the design right thorough; for if I do that I'm forcing the design down to an elementary level. If I am to break right through into a new world of performance I must find what my theory really needs, attend only to that, and then let the rest of the machine be assembled literally at random.[87]

However, once that approach was put into the metal, almost right from the start, the lack of recognition of an identifiable pattern begun to be gradually evident, acknowledging that "DAMS has reached the size of ten valves, and it has proved exceedingly difficult to understand".[88] Ashby patiently observed DAMS' behavior for two years, waiting to see if it would organize itself in an intelligible way. No pattern was to be found, the machine behaving just plainly chaotic, without achieving any stability by itself—ever.

It is almost painful as it is fascinating to witness how Ashby patiently tries to tinker with parts of the machine here and there, pulling it apart and re-assembling it again, so that some kind of pattern of stability would obtain—and the consequent ruminations that he has about the workings of the biological world after each tinkering.[89] It would show that his personal views on fundamental issues, such as Darwinian evolution, were being affected by it. For

---

[86] Ashby wrote in his diary that:

> . . . to find the form was difficult. I thought about a number of possibilities but found it difficult to satisfy myself that they were right; for the possibilities got so devilish complicated that I found them difficult to trace them right through from design to performance (Ashby 1928–1972, p. 2950).

[87] Ibid.

[88] Ashby 1928–1972, p. 3148.

[89] Pickering suggests that reconstructing the painstaking story of this machine would be a serious task worth pursuing (Pickering 2010, p. 110).

instance, within the mindset of "throw design to the winds",[90] in frustration with the crippling complexity of designing a working DAMS, he finds solace in methodologies friendly to such sort of approach:

> A good example of a system of knowledge, i.e. of action, that has been worked out almost entirely by Darwinian methods of trial, error, small improvements, and always following whatever seemed better is the technique of histological staining, especially as shown in the variety of methods worked out by Cajal and the Spanish school. Many of their methods were developed to a high degree of effectiveness, and without the slightest rational foundation.[91]

It became increasingly obvious for Ashby that certain "fixes", "ancillary regulations" had to be applied for the ultra-interconnected machine to show some kind of stable state. However, it would seem that one would be reverting back to the issue of "design". There is no way to account for the needed "gate" that would control the passage of information (the flow of electric current), whenever it is required, in a sufficiently complex process of adaptation. But this gate allegedly has to exist, since complex (biological) organisms in nature do adapt. The question regarding the entity that controls this gate seems to remain.[92] After all, the earlier homeostat's random search was also an issue for some of the cyberneticians, for whom the idea of order as naturally emerging out of chaos due to inherent properties of matter was an uneasy concept—and unfaithful to Wiener's understanding of entropy.[93] Ashby would later develop an answer that could be seen in stark contrast with the "throw design to the winds" quote above . . .

---

[90] Ashby 1928–1972, p. 2950 (quoted above).

[91] Ashby 1928–1972, pp. 3526–3527. Ashby is likely referring here to the "inductive" method (data collection-analysis-theory) used by Charles Darwin in his scientific practice, as opposed to the "deductive" one (theory-data collection-analysis), allegedly more faithful to a traditional understanding of the scientific method. For an accessible explanation of Darwin's inductive method within his theory of natural selection, see Lee 2000, ch. 2.

[92] The issue of "design" in nature seems to be far from settled, and it shall not come as a surprise that Ashby's name appears in contemporary treatments of the "science and religion" debate. See Huchingson 2005, p. 304; Rolston 2006, pp. 114–115.

[93] See Chapter 2, Section "The Macy Conferences".

The provision of the ancillary regulations thus demands that a process of selection, of appropriate intensity, exist. Where shall we find this process? The biologist, of course, can answer the question at once; for the work of the last century, and especially of the last thirty years, has demonstrated beyond dispute that natural, Darwinian, selection is responsible for all the selections shown so abundantly in the biological world. Ultimately, therefore, these ancillary regulations are to be attributed to natural selection... There is no other source.[94]

This biological "Maxwell's Demon" in charge of a necessary-for-survival gateway operation is, according to Ashby, "Darwinian evolution". As it is now known, Darwinism, beyond "natural selection", was eventually improved via Mendelian genetics. In fact, this improvement accounts for much of Darwinism's explanatory *mechanisms* that would account for the mutation of a surviving organism in the first place. In face of this, Ashby's radical *tabula rasa* epistemology (anti-nativist to the point of ascribing life characteristics to matter itself, and thus pushing the burden of causality to environmental influence alone) seems to collapse here. Evolution would have provided then these life-saving *homunculi* via genetic inheritance. There *is* something already furnished with nature that allows for living things to survive. If Ashby was not caught in a metaphysical web in conceding that the gateways would be the turf of "Mother Nature"(!) he did at least accept that whatever takes place *ad intra* the organism does complement behavioral cues. The "structure" of the living system, in particular its nervous system—and most interestingly, its brain—ultimately matters.

Models based upon measurable behavior are not enough past a certain point of complexity. After that point is reached, it would seem that chaos ensues. Ashby might have robbed cybernetics of the impetus that the enterprise had as one of its most important strengths for growth and development. But if he did that, he did it in accordance to cybernetic dogma and by actualizing a downward spiral that was already there in potential. One of these seeds pointed to a more troubling aspect of cybernetics—from the point of view of their scientific stand regarding studied phenomena and the

---

[94] Ashby 1960, pp. 229–230.

role of models in science. The early alert regarding this phenomenon was given, in fact, by von Neumann. The next chapter will show how those concerns were indeed pointing to features of cybernetics' own undeclared metaphysics that would eventually prove to be harmful for the consistency of the project. A treatment of that aspect of the cybernetic development toward its own collapse is the content of the next chapter.

CHAPTER 7

Emphasizing the Limits of a Machine Epistemology

John Von Neumann's Appropriation of McCulloch and Pitts' Networks

One of the fiercest proponents of cybernetics even before the time it became widely known, John von Neumann was a prominent Hungarian Jewish mathematician educated in an elite Protestant school. He fled the growing Nazi terror and landed in the USA, progressively becoming, probably partly due to émigré effusiveness, profoundly involved in "national identity building" military projects, such as the development of the atomic bomb.[1] The role of von Neumann in the cybernetic enterprise has been, up until now, regarded as somewhat secondary. Yet, he actively engaged in most of the Macy conferences, having kept close contact with its

[O]ne could draw within plausible time limitations a fictitious "nervous network" that can carry out all these functions... It is possible that it will prove to be too large to fit into the physical universe. (Von Neumann 1951, pp. 33–34).

---

[1] Having chosen Hiroshima and Nagasaki as the most convenient cities to drop the bomb, and tirelessly advocating a preventive nuclear strike against the Soviet Union, von Neumann had during his last days at the hospital, next to his bed, "the Secretary of Defense, the Deputy Secretary of Defense, the Secretaries of Air, Army, Navy and the chiefs of staff", asking for his last strategic advice (Rédei 2005, p. 7).

© The Author(s) 2017   171
A. Malapi-Nelson, *The Nature of the Machine and the Collapse of Cybernetics*, Palgrave Studies in the Future of Humanity and its Successors, DOI 10.1007/978-3-319-54517-2_7

main protagonists: Norbert Wiener, Warren McCulloch and Walter Pitts. In fact, the constructed computer machine whose internal design became known as the "von Neumann architecture" was directly inspired by cybernetic discussion.[2] Probably more importantly, some of his insights surmounted deep philosophical issues within cybernetics that might have not been fully grasped at the time by its own members, but that nevertheless could have contributed to cybernetics' own implosion.

Originally interested in quantum mechanics and in the incipient conceptualization of information in physics, von Neumann maintained a close friendship with Norbert Wiener from the mid-1930s on. As mentioned earlier, both Wiener's insights into information and entropy, and the philosophical and scientific consequences coming out of them, were very close to his inner interests. Indeed, he probably felt having been beaten by Wiener into showing first the import and relevance of these connections.[3] The beginning of the 1940s, however, marked a distinct moment in his intellectual journey. Having read Turing's paper "On Computable Numbers",[4] and while being deeply impressed by it, he still could not get around the idea of applying such an outlook to the human mind. For him[5] the apparent ontological rivalry between the rich, creative and messy human mind, versus the dry, exact and rigorous logic that came out of it, entailed an enduring interrogation mark. Indeed, he referred to Turing's brilliant paper, in his own words, as being very "anti-neurological"[6]—until he came across McCulloch and Pitts' article.[7] This paper, itself heavily influenced by Turing's insights, struck a chord in von Neumann for two important reasons. One of them, of a more practical nature: he needed a machine to perform massive number crunches in order to advance the secret development of a nuclear weapon. McCulloch and Pitts' paper, which in his own words was needlessly convoluted due to the use of Carnap's logical nomenclature, was nevertheless the basis for his

---

[2] Dupuy 2000, pp. 65–69.

[3] Chapter 3, Section "Traditional Explanations for the Collapse of Cybernetics".

[4] Turing 1937.

[5] As for Warren McCulloch (See Chapter 5, Section "Machines can be Immaterial").

[6] Von Neumann 2005, p. 278.

[7] McCulloch and Pitts 1943.

EDVAC project—what became the world's first stored-program computer, whose architecture still in use in most computers was referred above.[8]
Von Neumann' secret and active engagement in atomic weaponization
seemingly made of him a cyberneticist with his own non-disclosed agenda.
Although several cyberneticists were involved in military projects (cybernetics itself having come out of an effort for facilitating complex military
tasks), von Neumann went the extra mile in using cybernetic resources for
his own military endeavors. As previously mentioned, he made Wiener buy
into believing his willingness for coming to MIT to work with him,
provided there would be adequate funding for constructing a remarkably
expensive computing machine. Once the offer by MIT became concrete
(thanks to Wiener's intense lobbying), Princeton University, being von
Neumann's home institution, took serious notice, and in turn offered

[8] Chapter 5, Section "Machines Can Be Immaterial". Although beyond the scope
of this book, the popular ascription of "father of the computer" to von Neumann
is lately disputed mainly by British scholars who resent that Turing is not widely
acknowledged as responsible for todays' computers. The traditional claim is that
the American side, with all its resources, managed to "once again" kidnap an
original British idea, exploit it and reap its benefits. A report by Stanley Frankel,
a colleague of von Neumann at Los Alamos, is brought up, seemingly remarking
von Neumann's reliance on Turing's 1937 "On Computable Numbers". Indeed,
US mathematicians and engineers allegedly were "brute forcing", by means of
complicated and expensive hardware, the instantiation of von Neumann's fairly
vague ideas laced with biological terminology (more on this below). Also, the
engineer in charge of translating von Neumann's manuscript into a physical
machine, Harry Huskey, refers to it as being "of little help to those of us working
on the project." It is alleged that Turing's approach, in contrast, mindful of the
scarcity of material resources, would come up with clever solutions requiring the
least possible hardware. On the American side, a report by Julian Bigelow is
brought up, where he advances that the merit truly relies on von Neumann's
acumen to flesh out Turing's philosophical insights into the metal. In fact,
Turing's mentor, Max Newman, spoke of von Neumann as being an "applied
mathematician at heart"—despite the latter's brilliance in articulating theoretical
issues (e.g., in quantum physics, at the beginning of his career). Since both British
and American governments keep de-classifying WWII documents, the issue is far
from closed—in fact, it is gaining momentum among technology historians,
particularly from the British side. See Hodges 1983, ch. 5; Copeland 2004,
pp. 21–27; Huskey 2005, p. 283; Copeland 2012, ch. 8; Copeland and
Proudfoot 2005, pp. 112–117.

funding to build the machine there. Thus, von Neumann never went to MIT—as he probably never intended to do in the first place, merely using MIT's offer as negotiation leverage with Princeton. Furthermore, he even asked Wiener to send his own close collaborator, Julian Bigelow, to work with him—to which Wiener candidly agreed. And so Bigelow became chief engineer for the building of EDVAC.[9] All these moves were performed while portraying the building of the computer as instantiating a genuinely cybernetic aim—which in part it was. But only in part, as we will see. For the time being, the secret task of collaborating toward a nuclear weapon was conveniently left unspoken.

In the famous *First Draft*[10] for building the EDVAC, it was mentioned that more than one engineer found it odd that von Neumann used biological terms for referring to the parts of the future machine—such as neurons, synapses and organs.[11] This oddity points toward the less practical and more philosophical second reason behind von Neumann's deep appreciation of the McCulloch and Pitts' paper. He was impressed by the apparently successful effort of linking the two allegedly naturally incompatible entities—mind and logic, the latter having strangely come out of the former. The 1943 paper gave a mechanistic account of how a nervous system in general, and the mind in particular, might work.[12] This was no small feat. Von Neumann enthusiastically jumped into the cybernetic wagon and only three years later he undertook the honor of giving the first ever formal talk on cybernetics at the Macy conferences in March 1946.[13] At this opening event he recounted and explained how his computer machine became a reality, clearly acknowledging his

[9] These strategically motivated moves, which showed von Neumann's acute political wit, might have contributed later to tensions with Wiener, as depicted in Chapter 5, Section "Machines Can Be Immaterial".

[10] Von Neumann 1945.

[11] There are those who believe that keeping the biological vocabulary in an engineering paper allowed von Neumann to circumvent the veil of secrecy that would be otherwise put by the military—and so he would be able to share it, discuss it and eventually improve it. Let us not forget that this was a tool being built in order to expedite the construction of atomic weapons. See Tedre 2006, pp. 211–215.

[12] See Chapter 5, Section "Machines Can Be Immaterial".

[13] See Chapter 2, Section "Machines are Embodied Theories".

theoretical debt to McCulloch and Pitts' work (This was followed, within the same presentation, by Lorente de Nó's suggested application of the model to brain physiology).

However, after von Neumann's 1943 "epiphany", while working on the construction of the computing machine (with an undisclosed eye firmly set on developing the nuclear bomb), he remained genuinely curious regarding the actual correlation between the McCulloch and Pitts' model on the one hand and real biological substrata on the other. Von Neumann, as much as McCulloch and Pitts,[14] defended the heuristic value of axiomatization and the resulting use of idealized, black-boxed neurons, only reactive in a yes/no fashion. Armed with this methodology, von Neumann engaged in extensive connections with the biological and medical community, exchanging ideas and discussing views.[15] It did not take long for him to begin showing warning signs regarding the feasibility of the newly devised artificial networks in what pertains to brain matters—at the same time that the computer based upon that very model was being duplicated by the dozens (since he did not copyright the now legendary *First Draft*).[16]

Tracing von Neumann's gradual coming to terms with the extreme complexity of the structure they were bracketing off for the sake of theoretical operability—the nervous system and the brain—is historically fascinating. Below is a critical account of what progressively grew into a schism within the cybernetic movement—a schism which opened up hidden assumptions whose implications are arguably letting themselves be felt up to our times.

---

[14] In all fairness to Walter Pitts, having undertaken himself a journey that started with logic and became progressively close to biology, he was always aware and cautious of such idealizations. Dupuy remarks that in a "curious exchange of roles, while the logician Pitts abandoned formal models to pursue experimental research on the nervous system, the neurophysiologist McCulloch was to devote all his efforts to theory and logic" (Dupuy 2000, p. 137).

[15] Ironically, there are more records left of these biomedical meet ups, than those concerned with cybernetics. See Aspray 1990, pp. 298–301.

[16] Situation that occasioned a bitter rift among those who were involved in the EDVAC project. The American mathematician turned computer scientist Herman Goldstine (1913-2004) was one of them, and he recounted the story in his book (Goldstine 1993, pp. 211–224).

The Letter to Wiener: Post Hoc, Ergo Propter Hoc?

As already mentioned, two years before the *First Draft* for building EDVAC,[17] and enthused by the possibilities opened up by McCulloch and Pitts' networks, von Neumann actively engaged in dialogue with the biomedical community. By November 1946, toward the end of the very same year when he inaugurated the 1st Macy Conference with the first talk given in March, von Neumann wrote a revealing letter to Norbert Wiener. Its contents are of considerable importance if one takes it as a probe into what was going on in his mind regarding the feasibility of the cybernetic enterprise for modeling the brain. The letter starts with a dramatic tone, putting forward the issues that disturbed him. He then elaborates on them, in order to end up suggesting a change of focus for their future cybernetic endeavors—a change that he eventually undertook as a prime task of his intellectual journey, indeed until the end of his days, one decade later.

Von Neumann started off by acknowledging that some issues were deeply troubling him for the most part of the year. One could safely assume that he already was entertaining these unsettling thoughts *while* giving his presentation at the 1st Macy Conference, nine months earlier. Von Neumann mentioned his reticence for letting him (Wiener) know of these troubles up until then, due to the lack of a positive alternative—which he did introduce toward the end of the letter. The general problem is stated as cybernetics having chosen, in his view, probably the most difficult object of study: the brain. Cybernetics relied on theories which can certainly give some kind of explanation, in a substantially mechanical manner. That much it is granted. But the situation in which the whole enterprise lies now, he continues, is not better than the moment in which it started. In fact, he believed that "after the great positive contribution of Turing-cum-Pitts-and-McCulloch is assimilated, the situation is rather worse than better than before".[18] The cause for this rests *both* in the extreme complexity of the subject at hand, and the theoretical models that have been used.

---

[17] Von Neumann 1945.

[18] Von Neumann 2005, p. 278.

More specifically, von Neumann somewhat resented that McCulloch and Pitts' network provided results in such a self-contained, exact but general manner, that it left no way to confront it with the object that they were supposed to understand in the first place. If the model ends up enjoying a space beyond the realm of testability (and thus beyond scientific verifiability) it might still be heuristically good for something, but likely not for a nervous system. This other "something" was of course already having a physical instantiation at the hands of von Neumann. Indeed, the weakness in the model's refutability aspect did not remove from it its proved productive capacity, as the construction of his computing machine was amply evincing. However, once we step back and begin to really pay attention to the structure of the object that we wanted to originally parcel and explain, the complexity is such that one is shocked into a sort of epistemic paralysis:

> After these devastatingly general and positive results one is therefore thrown back on microwork and cytology—where one might have remained in the first place... Yet, when we are in that field, the complexity of the subject is overawing.[19]

Von Neumann gave picturesque analogies of the situation in which he felt they were while trying to study the brain with the best of the cybernetic constructs. For him to "understand the brain with neurological methods seems to me about as hopeful as to want to understand the ENIAC... with no methods of intervention more delicate than playing with a fire hose... or dropping cobblestones into the circuit".[20] He then proceeded to expand on the nature of the problem they were facing, which revealed a second discouraging aspect. If one is to scale down the choice of degree of complexity of the nervous system to study, say from a human one to that of an ant, much of what can be grasped by the McCulloch and Pitts' model goes down the drain, since such simplicity in turn comes packaged with several blind spots: "As the digital (neural) part simplifies, the analogy (humoral) part gets less accessible, the typical malfunctions less known, the subject less articulate, and our possibilities of communicating with it poorer and poorer in content".[21] However, thereafter he

---

[19] Ibid., p. 278.
[20] Ibid., p. 279.
[21] Ibid., p. 279.

tried to turn the negativity around and finally produced, all things considered, a radical, short and revealing suggestion, signaling the venue that cybernetics had to pursue in order to render more feasible results: "I feel that we have to turn to simpler systems".[22]

Von Neumann tried to expose the flawed nature of the argument behind the cybernetic need for studying only multi-cellular entities. The idea behind such argument was that since a neuron is just one cellular entity, only a large number of them networked together could render something meaningful. However, he pointed out to the complexity of that one cell, which shows reproductive features, and in some cases, it appears to be a perfectly self-contained living organism. Having in consideration these characteristics, which point out to a rich and likely relevant complexity in the unicellular organism, von Neumann cast a dark shadow upon McCulloch and Pitts' networks: "This in itself should make one suspicious in selecting the cells as the basic 'undefined' concepts of an axiomatism".[23] Without specifically mentioning either scientists or their model, the reference to McCulloch and Pitts' networks is clear: The "black-boxology"[24]

---

[22] Ibid., p. 279.

[23] Ibid., p. 279.

[24] This black box approach was present in cybernetic gatherings right from the start, and it played a role beyond a typical behavioristic metaphor. In fact, understanding an unknown "enemy machine" would use this heurism—all the more relevant due to the ongoing war. The first cybernetic gathering at Princeton, as recalled by Warren McCulloch, shows that von Neumann was quite comfortable with a black box approach, when the situation merited it...

> Lorente de Nó and I, as physiologists, were asked to consider the second of two hypothetical black boxes that the allies had liberated from the Germans. No one knew what they were supposed to do or how they were to do it. The rust box had been opened and exploded. Both had inputs and outputs, so labelled. The question was phrased unforgettably: "This is the enemy's machine. You always have to find out what it does and how it does it. What shall we do?" By the time the question had become that well defined, Norbert was snoring at the top of his lungs and his cigar ashes were falling on his stomach. But when Lorente and I had tried to answer, Norbert rose abruptly and said: "You could of course give it all possible sinusoidal frequencies one after the other and record the output, but it would be better to feed it noise—say white noise—you might call

that brackets the somatic aspect of the neuron for heuristic purposes might be paying a high price. So high indeed, that both the studied baby and the epistemic bathwater might be, in fact, being disposed of.

Furthermore, the whole issue was probably being approached backwards —or to be more precise, upside down. Von Neumann questioned the wisdom of targeting the object of study (particularly if it is regarded as a highly complex machine) from the perspective of understanding its finished "whole"—itself prohibitively sophisticated and already displaying a behavior whose complexity explodes off the charts. Common sense would indicate that the way to approximate such a complex object would be via its simpler, relatively isolated parts, and then reconstruct the entire machinery from the ground up. In his words, the venue currently pursued by cybernetics did not amount to a reasonable methodology of study—once again, due to the uniquely complex nature of its object of study:

> Consider, in any field of technology, the state of affairs which is characterized by the development of highly complex "Standard components", which are at the same time individualized, well suited to mass production, and (in spite of their "standard" character) well suited to purposive differentiation. This is clearly a late, highly developed style, and not the ideal one for a first approach of an outsider to the subject, for an effort towards understanding. For the purpose of understanding the subject, it is much better to study an earlier phase of its evolution preceding the development of this high standardization-with-differentiation. I.e. to study a phase in which these "Elegant" components do not yet appear. This is especially true, if there is reason to suspect already in the archaic stage mechanisms (or organisms) which exhibit the most specific traits of the simplest representatives of the above mentioned "late" stage.[25]

this a Rorschach." Before I could challenge his notion of a Rorschach, many engineers' voices broke in. Then, for the first time, I caught the sparkle in Johnny von Neumann's eye. I had never seen him before and I did not know who he was. He read my face like an open book. He knew that a stimulus for man or machine must be shaped to match nearly some of his feature filters, and that white noise would not do. There followed a wonderful duel: Norbert with an enormous club chasing Johnny, and Johnny with a rapier waltzing around Norbert—at the end of which they went to lunch arm in arm. (McCulloch 1974, p.11).

[25] Von Neumann 2005, pp. 279–280.

Moreover, von Neumann suggested that they move further "down" in what concerns the optimal selection of an object of study. In going after elementary parts to treat, once they are better understood, cybernetics could have a better grasp of how the whole thing works. This methodology, von Neumann hoped, would in turn become a firmer building block for the improved cybernetic endeavor.

Von Neumann specified that they could draw the line in terms of how far down they should go once they reached the realm inhabited by *virus*. He reminded Wiener that such entities do show features that one could reasonably ascribe to full-blown living organisms, given that they "are self-reproductive and they are able to orient themselves in an unorganized milieu, to move towards food, to appropriate it and use it".[26] Von Neumann confidently went on to express that "a 'true' understanding of these organisms may be the first relevant step forward and possibly the greatest step that may be at all required".[27] The word "true" was purposely put by him inside quotations marks, as he went on to specify what he understood by such term in this context. His clarification of what is "true" no longer rings as being in communion with the "behavior-with-a-purpose" cybernetic article of faith—namely, that something is a machine because it *behaves* machine-like. This last move in effect entailed an important epistemological displacement in cybernetic dogma. Now the cybernetic take on "true" (understanding) would approximate that which was strategically bracketed off precisely for the sake of theory extension in the first place:

> I would, however, put on "true" understanding the most stringent interpretation possible: That is, understanding the organism in the exacting sense in which one may want to understand a detailed drawing of a machine, i.e. finding out where every individual nut and bolt is located, etc. It seems to me that this is not at all hopeless . . . I suppose (without having done it) that if one counted rigorously the number of "elements" in a locomotive, one might also wind up in the high ten thousands. Consequently this is a degree of complexity which is not necessarily beyond human endurance.[28]

---

[26] Ibid., p. 280.

[27] Ibid.

[28] Ibid.

The suggestion of turning their focus into the structure of the object, into the stuff that makes up the actual body of the organism, probably had a bitter taste for Wiener. Avoiding opening and tinkering with the innards of the black box was to a considerable degree what helped launching the cybernetic totalizing worldview in the first place. Recall Wiener's struggle with a reductionist notion of "behavior" and his subsequent musings at expanding it.[29] What von Neumann is suggesting goes frontally against what Wiener had so effusively proposed in his 1943 founding paper. Toward the end, that article leaves no doubt regarding the more-than-heuristic role that a behavioristic approach has in the study of animal and machine as one sole entity—different in degree of complexity, not in kind.[30] Wiener maintained that

a uniform behavioristic analysis is applicable to both machines and living organisms, regardless of the complexity of the behavior. It has sometimes been stated that the designers of machines merely attempt to duplicate the performances of living organisms. This statement is uncritical. That the gross behavior of some machines should be similar to the reactions of organisms is not surprising. Animal behavior includes many varieties of all the possible modes of behavior and the machines devised so far have far from exhausted all those possible modes. There is, therefore a considerable overlap of the two realms of behavior...A further comparison of living organisms and machines leads to the following inferences. The methods of study for the two groups are at present similar. Whether they should always be the same may depend on whether or not there are one or more qualitatively distinct, unique characteristics present in one group and absent in the other. Such qualitative differences have not appeared so far.[31]

Furthermore, Ross Ashby, the theorizer of cybernetic fundamentals— always in intellectual sync with Wiener—defined cybernetics as the science that "treats...not things but ways of behaving. It does not ask 'what is this thing?' but 'what does it do?'...It is thus essentially

---

[29] While constructing his AA-Predictor (Chapter 2, Section "The AA-Predictor").
[30] For the ultimate consequences of this view to the cybernetic enterprise at large, see Chapter 8, Sections "Mechanizing Machine-Unfriendly Realms: Surreptitious Metaphysical Commitments Die Hard" & "Constructability as a Measure of Epistemic Success: The Intractability of a Highly Complex Model".
[31] Rosenblueth et al. 1943, p.22.

functional and behaviouristic".[32] And we saw how steadfastly holding to this tenet cost him a high price.[33] All in all, we can see that an enriched behavioristic approach laced cybernetics from the beginning to the end, both on the American and the British sides.

Let us recall that von Neumann's critical observations are coming the very same year he was giving the inaugural Macy conference presentation. Expectably, the letter was not sent without some prudent hesitation. He consulted with Walter Pitts before sending it off.[34] What exactly happened after it was delivered remains largely unknown. We do know that a meeting between Wiener and von Neumann likely took place after the letter, in December—but we do not know what transpired there. We also know that Norbert Wiener did not mention anything about this letter to anyone—not in writing nor orally—ever. However, as it has been mentioned, soon after people began noticing a discernible distance between these two hitherto close old friends. Was one fact caused by the other, given that one occurred after the other? Beyond merely correlating events, one can only speculate a causal connection between both. And even if the correlation seems to be fairly strong, we now know that there was more going on.[35] Still, there is a curious epilogue regarding this letter, which might partially explain von Neumann's posterior ardent impetus for further developing his critique against the (at the time) current methodological philosophy of cybernetics.

As mentioned earlier, from 1943 on, due to McCulloch and Pitts' bold hypothesis regarding the brain, von Neumann developed an active interest in biology and physiology, taking advantage of fruitful contacts in the biomedical community. In fact, McCulloch regarded von Neumann's collaborative labors as the reason why the biological community took interest in his paper. Less than a year after the controversial letter, at the 3rd Macy Conference,[36] von Neumann successfully lobbied to bring the German biophysicist Max Delbrück (1906–1981) as an invited guest. Von

---

[32] Ashby 1962, p. 2.

[33] That is, the case of DAMS (Chapter 6, Section "The Backlash: Unforeseen Consequences of a Behavior-based Ontology").

[34] Macrae 1992, p. 107.

[35] See Chapter 3, Section "Traditional Explanations for the Collapse of Cybernetics".

[36] March 1947.

Neumann had in fact mentioned Delbrück's work in embryology and bacteriophage viruses in the letter to Wiener. All the intentions were set in place for christening Delbrück as a permanent member of the cybernetic group. For von Neumann, Delbrück would be the perfect glove for the cybernetic hand: a physicist turned biologist, trying to scientifically (and thus, for them, mechanically) understand life, paying due attention to the most elementary building blocks of a living system. Once the invitation was officially arranged, the events that followed it took a somewhat bizarre turn.

Max Delbrück, now formally invited, could not attend the 4th Macy Conference, but he did make it to 5th one, in the Spring of 1948. It can safely be assumed that von Neumann was quite excited about this. After all, the scientist who symbolized the field of research he thought to be the next platform for cybernetics to take off from, was showing up at the cybernetic hub. As it turns out, right after the conference ended, Delbrück made it clear that he was never going to come back again. Delbrück was indeed so disappointed, that he was quoted some years later characterizing the said Macy conference as "vacuous in the extreme and positively inane".[37]

In all fairness, it might have been a case of spectacular bad timing. On the first day that Delbrück attended, practically the whole meeting was devoted to issues of language. He might have felt of it, literally, as a waste of his time. Delbrück was a genuine physicist turned biologist, accustomed to the rigors of hardcore physics[38]—despite the cybernetic mission, there never was a physicist in the group (besides von Foerster, the editor of the conferences, who obtained his doctorate just four years earlier). He might have been turned off by the purposely "flexible" language, which always tried to accommodate the vastly, even disparately different disciplines being loosely matched at the gatherings—from mathematics and engineering to psychology and anthropology.[39] Or it could be that both reasons perfectly conflated into the disastrous (for von Neumann) outcome. Be that as it may, it is impossible to know with exactitude what kind of effect this reaction had on von Neumann. But one can safely assume

[37] As personally conveyed to Steve Heims years later (Heims 1991, p. 95).

[38] More on Max Delbrück's subsequent in roads in Chapter 8, Section ". The Relevance of Materiality for the Nature of a Machine: From Complexity to Disembodiment".

[39] Dupuy 2000, pp. 76–80.

that he did not take it lightly. The subsequent increasing sharpness of von Neumann's criticism against cybernetics can further fuel that assumption.

THE HIXON SYMPOSIUM: STRENGTHS AND WEAKNESSES OF THE MCCULLOCH AND PITTS' NETWORKS

Only three months later, von Neumann was again standing behind a podium giving a talk on cybernetics, this time at the famed Hixon Symposium.[40] By this time, two years after the infamous letter, he had his plan for the future development of cybernetics more in focus. Von Neumann took particular interest in the logic behind features that portray life in "very simple" organisms. A feature he was particularly interested in was reproduction. It was his understanding that thanks to Turing's "On Computable Numbers" it was no longer a mystery that a more complex entity can be begotten from a less complex one—something which apparently goes against common sense, but that nevertheless is allegedly a key feature of evolution. As von Neumann read Turing, a universal Turing machine embedded with sufficient instructions can produce (the behavior of) any other machine, provided that rigorous and unambiguous instructions are furnished. Von Neumann remarked that this is no more mysterious than the capacity we have for reading pretty much anything in a certain language after its rules of grammar are firmly entrenched in our cognitive system.

However, the logic required for the reproduction of artificial entities would have to be attentive to details pertaining to their structure and its elements. Such future logic should take into consideration issues of heat, material fatigue, friction, and so on. It would be a logic acquainted with issues of thermodynamics. That much von Neumann had clear by that time. And the way he introduced "The General and Logical Theory of Automata"[41] was by means of pointing out the deficiencies of current cybernetic models. There had to be, after all, a core reason why cyberneticists should turn their attention to his new proposal.[42] Hence, the McCulloch and Pitts' model received

---

[40] September 1948. Published as Jeffress 1951.

[41] Von Neumann 1951.

[42] Later on von Neumann regarded his subsequent work in theory of automata as going beyond being a solution for an ill-oriented cybernetics. He saw his

a gloves-off attack whose entire array of consequences might have not been thoroughly evaluated until these days.[43]

Warren McCulloch might have already been aware of von Neumann's both positive and critical attitude regarding his (and Pitts') model. It can be safely assumed that by this time he understood well that von Neumann was cheerful about his model only in so far as it opened up possibilities for constructing computing machines based upon it—machines that would perform certain tasks that until that moment where done by humans,[44] but incomparably faster, and basically flawlessly. McCulloch likely was indeed aware that von Neumann was however less than happy, in fact probably disturbed, by the association of such model to an actual brain or nervous system.

McCulloch gave a talk at the very same conference entitled "Why the Mind is in the Head".[45] There he made a reference to von Neumann's computers, somewhat jokingly pointing to their inherent limitations. In praising the practical efficiency of the brain's neural networks, McCulloch said that if von Neumann's machines would attempt to come any closer to the overall behavior of the brain, then the kind of resources demanded would exceed current practical feasibility...

developing theory as a future cornerstone for the revision of not only thermo-dynamics, but of logic and mathematics in general. Thus the theory of automata was meant to be his most important legacy—correspondingly speaking of it with solemnity until the moment of his death, in 1957. That might have been the reason why he chose to pursue it alone, unlike any other project he was involved with up to that point. See Shannon 1958, p. 123; Von Neumann and Burks 1966, p. 28; Goldstine 1993, p. 285.

[43] Some people besides McCulloch did not like the criticism. Marvin Minsky, a doctoral student of von Neumann and later pioneer of AI, did not regard it well. Although the newly formed wave of AI researchers saw symbolic logic as superior to artificial neural networks, they were unfriendly to the idea that neural networks (and probably by extension, their own symbolic models) were that far removed from what actually occurs in the brain (See Dupuy 2000, pp. 65–69).

[44] Called computors, or number crunchers—a remarkably tedious job. See Chapter 4, Section "Alan Turing's Strange Reversal Regarding the Question of Artificial Intelligence".

[45] McCulloch 1951.

Neurons are cheap and plentiful. If it cost a million dollars to beget a man, one neuron would not cost a mill. They operate with comparatively little energy. The heat generated raises the blood in passage about half a degree, and the flow is half a liter per minute, only a quarter of a kilogram calorie per minute for 10, that is, 10 billion neurons. Von Neumann would be happy to have their like for the same cost in his robots. His vacuum tubes can work a thousand times as fast as neurons, so he could match a human brain with 10 million tubes; but it would take Niagara Falls to supply the current and the Niagara River to carry away the heat.[46]

These extraordinarily problematic outcomes would be happily acknowledged by von Neumann, with the following twist. These problems are in fact expectable, since McCulloch and Pitts' artificial networks are good for constructing computing machines, not for telling us how our nervous system, in fact *any* nervous system, works. Further, von Neumann went the extra mile in the discussion after his own talk at the conference, pointing to even more mind-blowing outcomes if one is to attempt modeling the nervous system with such networks. The problem is not his machines: the problem is the model.

Von Neumann started by pointing out the methodology that validated the formal networks in the first place. Faced with the extreme complexity characteristic of natural organisms, scientists can be justified (given their mechanistic presupositions) in understanding the object of study as constituted by smaller, fairly isolated parts. This view in turn can subdivide the problem in two sub-tasks: 1) The attempt to understand "the structure and functioning of such elementary units, individually"[47] and 2) "Understanding how these elements are organized into a whole, and how the functioning of the whole is expressed in terms of these elements".[48]

The first task was dismissed by von Neumann as extremely (sometimes impossibly) demanding, being the core of nitty-gritty scientific experimentation, in which humans will engage for likely a long time to

[46] McCulloch 1951, p. 54.
[47] Von Neumann 1951, p. 2.
[48] Ibid.

come.[49] He is concerned with the second task. A person with mathematical or logical inclinations can find fertile land here. How? By means of "axiomatizing", putting into "black boxes" all the chemico-biological elements underlying the basic units of the organic system—in this case, the neuron. The job concerning what takes place *ad intra* these heuristically convenient brackets is left to the laboratory science practitioner underdog—task 1), that is.

Once this convention is put in place, the work gets greatly simplified. We assume of these "neurons" as being "automatisms, the inner structure of which need not be disclosed, but which are assumed to react to certain unambiguously defined stimuli, by certain unambiguously defined responses".[50] These basic units admit only of a yes/no type of reaction, allegedly being the biological instantiation of a logical gateway, and thus amenable of being mapped upon good old logical connectives (not, and, or, if... then). The result is that "any functioning in this sense which can be defined at all logically, strictly and unambiguously in a finite number of words can also be realized by such formal neural networks".[51] Von Neumann acknowledged that the implications of this methodological "move" are as tremendous as they are admirable. In a sentence that resembles one of Ross Ashby's ultra-cybernetic lines of reasoning, he evaluated and appreciated it thus...

> McCulloch and Pitts' important result is that any functioning in this sense which can be defined at all logically, strictly, and unambiguously in a finite number of words can also be realized by such a formal neural network. It is well to pause at this point and to consider what the implications are. It has often been claimed that the activities and functions of the human nervous system are so complicated that no ordinary mechanism could possibly perform them. It has also been attempted to name specific functions which by their nature exhibit this limitation. It has been attempted to show that such specific functions, logically, completely described, are per se unable of mechanical, neural realization. The McCulloch-Pitts result puts an end to

---

[49] Note that this is precisely what von Neumann suggested Wiener to do in the letter, two years earlier (Section "A Machinal Understanding of an Algorithm and the Material Liberation of the Machine", above). More on this reversal below.

[50] Von Neumann 1951, p. 2.

[51] Ibid., p. 22.

this. It proves that anything that can be exhaustively and unambiguously described, anything that can be completely and unambiguously put into words, is ipso facto realizable by a suitable finite neural network.[52]

There would finally be a glimpse of an answer to the age-old question regarding how both exact disciplines and rigorous thinking can emerge from heretofore the antithesis of mechanical exactitude: the mind. This was not only the question that kept Warren McCulloch out of the Theology Seminary. It also made von Neumann plunge into the cybernetic project, later coming out with a full-blown computer. In fact, this insight was probably the one that, along with Wiener's views on teleology, drove scientists of the time into the cybernetic movement in the first place.

However, that was as far as von Neumann would go in terms of praising the cash value and heuristic justification of the neural model. Subsequently, he picked up where he left off regarding such a remarkable achievement—namely, making every behavior that can be rigorously and unambiguously described, feasible of being run by these networks. Von Neumann began to explore the downside of this methodological mandate. Specifically, he remarked on two basic questions that emerge: 1) "whether that network can be realized within a practical size, specifically, whether it will fit into the physical limitations of the organism in question"[53] and 2) "whether every existing mode of behavior can really be put completely and unambiguously into words".[54]

Regarding the first one, once again von Neumann left it as the stuff for painstaking hard science tinkering: task 1), above—however, he would later make a clear reference to this issue within the discussion that followed suit right after his presentation.[55] Regarding the second question, he tackled the issue by means of an example. What would it take to understand how we can know a triangle? Prima facie, applying McCulloch and Pitts' networks we can describe clearly the characteristics of a triangle (3 sides, 180 degrees). However, we soon reckon that we are also able to recognize circular triangles, deformed triangles, upside down triangular figures, and so on. Suddenly, we realize that the more accurate our

---

[52] Von Neumann 1951, p. 23.
[53] Ibid.
[54] Ibid.
[55] Ibid., pp. 32–41.

description is, the lengthier our report becomes. Nevertheless, even if it gets very long, it is still feasible ... That is, until we clash into a second aspect of the issue, the tractability chances of which look as bleak as they look discouraging:

> All of this [the set of descriptions], however, constitutes only a small frag-
> ment of the general concept of identification of analogous geometrical
> entities. This in turn is only a microscopic piece of the general concept of
> analogy.[56]

Von Neumann questioned the plausibility, in terms of practicality, of totally and finally describing the behavior of an organ as the good route to take—even if it is feasible in principle. Here he re-introduced what was advanced to Wiener in his letter two years before: the very un-cybernetic idea of opening the "black box" and describing the actual "connections of the visual brain".[57] In what amounts to a radical change of modeling strategy, von Neumann suggested swapping the model for the modeled as the most realistic avenue to pursue, given the explosively unworkable nature of describing all possible behaviors:

> It is not at all certain that in this domain a real object might not constitute
> the simplest description of itself, that is, any attempt to describe it by the
> usual literary or formal-logical method may lead to something less manage-
> able and more involved.[58]

After a certain threshold of complexity in the studied organ, it becomes more feasible to directly describe the very structure of the object, rather than accounting for all the possible behaviors of the said organ. This renders McCulloch and Pitts' networks good for very simple organisms, but utterly inadequate for something whose degree of complexity will entail that a constructed model would require the dynamic and cooling forces of the Niagara Falls—as McCulloch sarcas-tically suggested. Von Neumann did not spare any opportunity to clearly mention the "McCulloch-Pitts results" as no longer useful

---

[56] Ibid., p. 24.
[57] Ibid.
[58] Ibid.

when we attempt to describe a more complex organ, especially the brain—the most complex of them all. The discussion that ensued expectably had Warren McCulloch as the first respondent. What he said within his response was a hard confession that points to an acknowledgment of the critique:

> I taught neuro-anatomy while I was in medical school, but until the last year or two I have not been in a position to ask any neuroanatomist for the precise detail of any structure. I had no physiological excuse for wanting that kind of information. Now we are beginning to need it.[59]

After the behavioral neurophysiologist Ralph Gerard (1900–1974) asked for further explanations regarding the issue of rigorously describing complex behavior in a model, von Neumann took the occasion to relentlessly continue his critique. He accentuated the very practical problem of having to be able, when dealing with a complex object, to come up with a description that at least "can be read in a life-time".[60] Furthermore, in connection with the two first questions "dismissed" above,[61] and metaphorically vastly surpassing McCulloch's Niagara Falls' sarcastic reference, von Neumann asserted that using McCulloch and Pitts' networks to map the human nervous system would "turn out to be much larger than the one we actually possess".[62] *Much larger* is here used as a charitable understatement, as he then specifies that "it is possible that it will prove to be too large to fit into the physical universe".[63] Naturally, von Neumann consequently asked "What then? Haven't we lost the true problem in the process?"[64]

As we can see, both "first questions", initially dismissed as the stuff pertaining to the typical tasks of the hard sciences, returned in full force, since the consequence of modeling an extremely complex structure upon the

---

[59] Ibid.

[60] Ibid., p. 33.

[61] Regarding the "structure" of a studied object. See previous Section "Alan Turing's Strange Reversal Regarding the Question of Artificial Intelligence".

[62] Ibid., p. 34.

[63] Ibid.

[64] Ibid.

possible behavior of such structure incurs into a growth explosion in terms of model size. Once the threshold of becoming factually non-manipulable is reached, we seem to have betrayed the whole point of the endeavor, namely, attempting to understand the organism. This reversal might have been profoundly influential for cybernetics' own fate, instantiated in implications that failed to be recognized in full. A direct outcome, however, is below.

The Illinois Lectures and the Kinematic Model: Cybernetic Materialization of a Theory

Von Neumann reassumed the attack during a series of five lectures he gave at the University of Illinois in March 1949.[65] It is relevant to point out, however, that just as in the Hixon Symposium, the criticism of McCulloch

---

[65] Nine months earlier (and six months after the Hixon Symposium), at the Sixth Macy conference in March 1949, no one other than Warren McCulloch was in charge of delivering at the opening presentation a message from von Neumann— who could not attend. McCulloch faithfully conveyed what he called "von Neumann's warning" to the audience. The gist of the warning had to do with suspected inadequacy of taking for granted that 10^{10} neurons is not enough for accounting for the human brain. A signaling fact toward that assessment would be that the mere 300 neurons of the ant cannot afford to explain its behavior. There were some expressions of discomfort regarding such change in spirit: 10^{10} was regarded as an adequate estimate for neuronal population just at the previous (Fifth) Macy conference.

It became gradually apparent that there was a confusion between synapses and actual neurons, made evident by the parallel confusion regarding whether to ascribe the analogy of computer's vacuum tubes to synapses or neurons. Each analogy relation would of course lead to different outcomes. Walter Pitts finally brought the discussion to a halt, proposing that von Neumann got his calculations wrong: Seemingly von Neumann was referring to synapses (although McCulloch did mention "neurons"). A neuron has 2^n firing possibilities. "Therefore, 2^n, n is the number of distinct afferents which go to it, is the maximum. I don't see how you can get up to 1010 in a possible cell." That is how the artificial networks were conceived in the first place. It would also seem to follow that it is safe to assume that ants in fact have more than 300 "thinking" neurons.

This minor bleep did not stop von Neumann—it merely stopped *that* discussion. He remained undeterred, as we will see below (Von Foerster, Mead and Teuber 1950, p. 17).

and Pitts' networks was not the point of his presentation. Rather, the criticism was introduced as the point of departure upon which his "Theory of Self-Reproducing Automata"[66] would soon be developed, advancing it as "the way to go" precisely because of the unfeasibility of using McCulloch and Pitts' networks for modeling very complex organisms. The second of the five lectures is the one where von Neumann addressed McCulloch and Pitts' networks squarely, in the introductory manner just indicated. In this lecture, he gave some more details pointing to the inadequacy of the model in what pertains to the modeling of the brain. Von Neumann would later on be more specific about what is being lost if one is to choose this neural model as the instrument for studying the nervous system. The line of criticism was consistent with his previous remarks on the issue.

Von Neumann started off discreetly, just letting slip at this point that he is developing a slightly dismissive stance regarding the implication of having McCulloch and Pitts' networks as a plausible counterpart for the nervous system's natural networks. He expressed that their neural model "has a meaning which concerns me at this moment a little less...They wanted to discuss neurons".[67] The way McCulloch and Pitts proceed to go about this, von Neumann continued, is by means of using what is "known in mathematics as the axiomatic method, stating a few simple postulates and not being concerned with how nature manages to achieve such a gadget".[68] But he to some extent justified this strategy, pointing out the already mentioned remarkable achievement of showing how anything amenable to rigorous description can be instantiated by the model. Indeed, von Neumann cautiously endorsed the heuristic simplification attained after the encapsulating power of such idealized neurons, given that "one gets a quick understanding of a part of the subject by making this idealization".[69]

[66] Published posthumously, almost a decade after von Neumann's untimely death in 1957, as Von Neumann 1966. His student Arthur Burks took the task of gathering, heavily editing and even completing von Neumann's manuscripts on his theory of automata.

[67] Von Neumann 1966, p. 43.

[68] Ibid.

[69] Ibid., p. 44.

But then von Neumann, in stronger language, started to gradually convey the problems being brought up as side effects of such accomplishment. He asserted that the scientific duo "believed that the extremely amputated, simplified, idealized object which they axiomatized possessed the essential traits of the neuron, and that all else are incidental complications which in a first analysis are better forgotten".[70] Von Neumann wondered whether this move—namely, leaving aside the details that constitute a real neuron, for the sake of the model's operability—would ever reach wide acceptance. He doubted it.

Still in another instance, displaying a half-sympathy for the axiomatic strategy, von Neumann went on to describe a neuron as having "two states: It's excited or not. As to what excitation is, one need not tell".[71] In what seems as approvingly allowing for the bracketing of an otherwise important detail, von Neumann faithfully described McCulloch and Pitts' neuron, specifying that "its main characteristic is its operational characteristic and that has a certain circularity about it: its main trait is that it can excite other neurons".[72]

The last nod of approval toward McCulloch and Pitts' networks is given right before von Neumann begins to point out its shortcomings—although from a different angle than what was indicated at the Hixon Symposium. Graciously mentioning yet again the rigorous descriptive parsing accomplished by the model, he followed it with still another benefit, not without remarking that these outcomes carry a philosophical twist:

> May I point out what follows from this from a philosophical point of view, and what does not follow. It certainly follows that anything that you can describe in words can also be done with the neuron method. And it follows that the nerves need not be supernaturally clever or complicated. In fact, they needn't be quite as clever and complicated as they are in reality, because an object which is a considerably amputated and emasculated neuron, which has many fewer attributes and responds in a much more schematic manner than a neuron, already can do everything you can think up.[73]

---

[70] Ibid.
[71] Ibid.
[72] Ibid.
[73] Ibid., p. 46.

Now the requirement of intellectual honesty had been fulfilled, having given the best possible depiction of the model "on trial". Von Neumann went on to begin listing, in an almost accusative tone, the high prices that were being paid for the theoretical package purchased—and the great losses obtained in return. In his words, next to the great benefits attained by the McCulloch and Pitts' networks, "what is not demonstrated by the McCulloch and Pitts result is equally important".[74] There are three major issues left unsolved, or even further problematized, after assuming the McCulloch and Pitts' networks. What von Neumann recounted here summarized the discomfort he felt for three years: 1) There is no proof anywhere that McCulloch and Pitts' networks actually occur in nature.[75] 2) There is no assurance that whatever was left aside in the axiomatization of the neuron is not of the utmost importance. 3) Most disturbingly. . .

> It does not follow that there is a considerable problem left just in saying what you think is to be described. Let me try to put this in another way. If you consider certain activities of the human nervous system, you find that some of them are such that all parts of them can be described, but one is flabbergasted by the totality of what has to be described.[76]

Once he justified plenty why McCulloch and Pitts' networks were not the way to go for a cybernetic articulation in general—and of a nervous system

---

[74] Ibid.

[75] This points out to the devastatingly simple question, relevant these days (e.g., in cognitive science), whether a proposed model, whichever it is, actually *occurs* (namely, it is at least part of a physically instantiated mechanism). It may be argued that despite the simplicity and relevancy of the inquiry, its sole mentioning generally fosters confusion. The question about the ontological status of models is one that has largely gone underground. This issue is not unrelated to the one dealing with entities sharing a common fundamental trait, differentiated quantitatively or qualitatively. For the relation between the model and the modeled in what concerns a) the fate of cybernetics, see Chapter 8, Section "Constructability as a Measure of Epistemic Success: The Intractability of a Highly Complex Model"; b) a historical-philosophical precedent, see Chapter 9, Section "Viconian Constructability as a Criterion of Truth"; c) a contemporary development in science, see Chapter 10, Section "The Relevance of Materiality for the Nature of an Object: Simulation as Scientific *Factum*".

[76] Von Neumann 1966, p. 46.

in particular—von Neumann advocated for the line of research proposed to Wiener in the 1946 letter. Primitive organic systems should be the first ones to be understood, putting special attention on bacteriophages. The feature of primitive life that one ought to focus on is self-reproduction. If one is able to recreate such an essentially biological feature, von Neumann would consider it the first qualitatively distinct step toward a practical understanding of a living system feasibly reproducible in a model. Recreating self-reproduction would be a concrete task that could give a distinctly profound insight into the nature of life. In faithfully cybernetic manner, he thought of a real-world physical model of this primitive system: he specifically called for an automaton occupying Euclidean three-dimensional space. The primitive automaton would arbitrarily measure some decimeters, and it would be constituted by the following elements:

+ Four computing elements (three logical and one processing):
- Disjunctive organ (for the function *or*).
- Conjunctive organ (for the function *and*).
- Inhibitory organ (for the function *not and*).
- Sensing element (stimuli producing).
+ Four mechanical elements.
- A fusing element; this organ would weld parts together.
- A cutting element; this organ would separate parts.
- A muscle organ; this organ would provide motion.
- A rigid organ; which would just provide structural rigidity.[77]

---

[77] Von Neumann gathered a small group of people at the Institute for Advanced Studies at Princeton University, likely as a preparation for his presentation at the Hixon Symposium. What was said in those talks, unfortunately never recorded or published, gave the most details regarding his early theory of automata—the "kinematic model", as Arthur Burks baptized it. Burks attempted to reconstruct what transpired there. In terms of the "elements" of the self-reproducing automaton, von Neumann reportedly named them as follows:

A *stimulus organ* receives and transmits stimuli; it receives them disjunctively, that is, it realizes the truth-function "p or q." A *coincidence organ* realizes the truth-function "*p* and *q*." An *inhibitory organ* realizes the truth-function "*p* and not-*q*." A *stimuli producer* serves as a source of stimuli. The

The automaton would operate in the following way. The "mother" automaton, in possession of the whole set of instructions for a model of itself, is to be left freely floating in a liquid substance, surrounded by physical parts of itself, freely floating as well. When the automaton would bump into a part, tests would be performed within itself in order to identify what kind of part it is (a muscle contracts, a rigid bar does not change, etc.). It was not specified how this contact process would precisely operate, but von Neumann suggested that there might be a "sensing" device, actuated by the muscle element. The muscle organ would also provide motion to the fusing and cutting organs. The automaton would gradually put all the parts welded together, and as the last step, it would transfer the set of instructions of itself,[78] for the "daughter" automaton to continue the reproductive process. In such way, the "miracle" of life would be a step closer to be fully artificially recreated in the metal, thus

fifth part is a *rigid member*, from which a rigid frame for an automaton can be constructed…These connections are made by a *fusing organ* which, when stimulated, welds or solders two parts together…Connections may be broken by a *cutting organ*, which, when stimulated, unsolders a connection. The eight part is a *muscle*, used toproduce motion (Italics original. Von Neumann 1966, p. 81. See also; Burks 1969, pp. 2–3).

[78] Let us recall Turing's efforts to make explicit the notion of an algorithm by means of rigorously describing its behavior—which facilitated the Turing machine (Chapter 4, Section "A Machinal Understanding of an Algorithm and the Material Liberation of the Machine."). Von Neumann recognizes here the value of Turing's call for rigorously and completely describing parts—that is, the elements of his Turing machine—later embraced by McCulloch and positively recognized by himself (Section "Alan Turing's Strange Reversal Regarding the Question of Artificial Intelligence.", above). In order to circumvent the problem of a machine abstracting a pattern for the structure of the "baby" machine, it would be more efficient to just transmit the clear set of instructions:

In any conceivable method invented by man, an automaton which produces an object by copying a pattern will go first from the pattern to a description and then from the description to the object. It first abstracts what the thing is like and then it carries it out. It's therefore simpler not to extract from a real object its definition, but to start from the definition. To proceed in this matter one must have axiomatic descriptions of automata. You see, I'm coming quite close to Turing's trick with

providing the kind of cybernetic true knowledge only possible when breaking down the problem into an engineering question.

A number of issues can readily be identified in the previous description of artificial self-reproduction. Firstly, von Neumann conspicuously avoided touching upon the issue of an energy source—all the more telling, after advocating for a future logic that should be sensitive to issues of thermodynamics (the amount of steps to reach an answer would have practical consequences in friction, decay, erosion, heat dissipation, etc.). Probably, he was going to add this feature later.[79] However, it is already signaling the strategy that he is going to be using: axiomatizing elements and black-boxing issues that otherwise would throw paralyzing obstacles for the experimental research to continue. After all, he did justify the same strategy used by McCulloch and Pitts' networks to *some* extent.

Also, there is no elaboration as to how the grasping of the floating parts would be performed. Presumably by means of a movable mechanical "arm", helped by both the muscle organ and the energy source, the part would be grabbed and put in the right spot. The same goes for the welding and cutting procedure; there is no mention of the way in which these would work. Or even for the all-important gadget that would function as a "recognizing device", which would presumably provide information for the proper accommodation of the found part. There is practically nothing said about this.[80]

universal automata, which also started with a general formal description of automata (Von Neumann 1966, p. 83).

[79] Such is Arthur Burks' speculation (Von Neumann 1966, p. 82).

[80] This is how von Neumann would scantily describe the self-reproduction of the automaton, to the best recollection of Burks:

> The constructing automaton floats on a surface, surrounded by an unlimited supply of parts. The constructing automaton contains in its memory a description of the automaton to be constructed. Operating under the direction of this description, it picks up the parts it needs and assembles them into the desired automaton. To do this, it must contain a device which catches and identifies the parts that come in contact with it... Two stimulus units protrude from the constructing automaton. When a part touches them tests can be made to see what kind of part it is (Von Neumann 1966, p. 82).

THE THEORY OF AUTOMATA: FROM EMBODIMENT TO ABSTRACTION

Despite the cautious justification for obviating these empirical aspects for the sake of the advancement of the theory, one can safely assume that this might have been a cold comfort for von Neumann. He was not only keenly aware of the cybernetic tradition of demonstrating knowledge by means of "recreating" it in the metal, as showed by Grey's tortoises, Shannon's rat, Ashby's homeostat and even the early Wiener's uncanny AA predictor. Von Neumann demonstrated a personal leniency, unlike many mathematicians, to implement his ideas in the flesh, being the primal example what came to be known as the von Neumann architecture computer.[81] Leaving aside the truly mechanical aspects of the automaton was likely not something he cherished. In fact, it is recalled that he would be seen walking around Princeton University with a smiley face and big boxes filled with Meccano pieces of tinker toys.[82] After all von Neumann did have the cybernetic mandate of knowledge as recreating as an integral part of his epistemology.

Conversation with colleagues gradually convinced him to leave aside the "problematic" approach to knowledge-gathering having physical substrata as a grounding factor. The Polish mathematician Stanislaw Ulam (1909–1984) persuaded him to leave aside bodily three-dimensionality in order to re-state the problem in a two-dimensional plane. Being an outstanding mathematician himself, von Neumann acknowledged that his three-dimensional "kinematic" model lacked the

---

Arthur Burks later attempted to explain how this might work, admittedly infusing this explanation with his own educated guess for what von Neumann might have had in mind (Burks 1970, pp. 3–6).

[81] See the note above (Section "John Von Neumann's Appropriation of McCulloch and Pitts' Networks") on the dispute regarding calling von Neumann "the father of the computer".

[82] Herman Goldstine recalls a happy-faced von Neumann buying the biggest set of Meccano toys available, and enjoying tinkering with them at the university. Once his (probably only true) friend Stanislaw Ulam convinced him of the mathematical advantage of switching his model to a two-plane reality, he gave away the tinker toys to Oskar Morgenstern's son (Goldstine 1993, p. 278).

mathematical "elegance" that might be required for future generations to build upon it. With the new proposal, void of a balancing physical checkpoint, it was expected that he would be able to give attention to the elaboration of a logic of reproduction that he claimed was not yet existent.

What von Neumann developed after the suggestion of his colleague is what came to be known as the theory of cellular automata.[83] It consisted of two-dimensional squares, called lattices, which would—under specific instructions based upon the condition of the contiguous squares (occupied or not)—"grow" and "reproduce". These two-plane automata would show a fairly complex behavior stemming from simple rules, which would allegedly fulfill the evolutionary evidence of complex structures coming from simpler ones (as it is shown in the biological reproduction of entities, commencing in unicellular beings). All this was to occur on a flat geometrical plane, completely void of Euclidean mass and thus free from the extra—implementationally and operationally "crippling"—complications of the three-dimensional world.

The details of this historic theory are not the concern of the present work. What is important to remark is the effective outcome of von Neumann's change of methodology. As indicated above, von Neumann thought of his later work as the most fundamental of his whole career. He used to talk about it with solemnity, effectively devoting to it the last decade of his life—to the chagrin of government officials and colleagues who wanted him more involved in work with computing machines. However, despite the effort of a few disciples,[84] the theory of cellular automata fell into oblivion[85] after

---

[83] Or just the "theory of automata," since the kinematic model was, as we saw, discarded. This new two-dimensional approach is dealt with in the second part of von Neumann's posthumous book (Von Neumann 1966).

[84] Mainly Arthur Burks. He published about von Neumann's automata theory in Burks 1959; 1960; 1963a; 1963b, 1966, 1970; 1974; 1975 and of course Von Neumann 1966, where Burks did the heavy editing and completion.

[85] Boden 2006, pp. 890–892.

von Neumann died.[86] It was attempted to be rescued[87] but it remains an effort whose punchline never arrived.[88]

Most importantly perhaps, cybernetics missed a fundamental arm in its own body of theoretical tenets: the utmost importance of the physical instantiations of their theoretical pursuits. Von Neumann effectively ended up taking the same path that he criticized just a few years earlier, provoking a focus in the abstracted model as if it would be the object of study itself. The importance of the structure of the object was shown to be particularly relevant and helpful when the mapping of possible behavior became extremely complex and beyond feasible management in practical terms. But once one takes a peek into the structure itself, the difficulty leaves us "flabbergasted"—as von Neumann would like to refer to it. And so the nuanced epistemic trip taken by von Neumann followed these stages: 1. the study of virus; 2. the attempt to recreate "primitive" self-reproduction; 3. the paralyzing effect due to the extreme complexity of the physical details of its mechanical model; and 4. the retreat into a non-physical realm. And this last one is precisely what happened to McCulloch

---

[86] Von Neumann reverted to the Catholic faith that he adopted when he married his first wife. He was accompanied by a Benedictine priest during the last days of his life, at the end of which he received the Last Rites (McRae 1992, pp. 378–379).

[87] Probably the most serious attempt was NASA's commission to study the feasibility of constructing and using self-replicating machines for space exploration (See Freitas and Gilbreath 1980). This technical report has a section on asteroid mining that has lately regained some momentum, due to the renewed interest in such potentially pecuniary enterprise (See Freitas and Merkle 2004).

[88] The relatively novel field of Artificial Life (ALife) does not take sides with the kinematic model as opposed to the cellular model, but it does acknowledge the former as the first model that fully grasped the essence of ALife—a fortiori, a successfully self-reproducing automaton would qualify as *living* for this discipline (Langton 1989). Theorizers of nanotechnology do recognize von Neumann's kinematic model as conforming the root of their envisaged "molecular assembler" (See Drexler 1986, ch. 4). For contemporary efforts regarding ALife and nano-technology, see Chapter 9, Sections "Transhumanism: An Age-old Grand Project for Humanity's Uplifting" & "Scotian Continuum between the Material and the Immaterial" and Chapter 10, Sections "Constructability as a Measure of Epistemic Success: Molecular Reordering of Reality" & "The Relevance of Materiality for the Nature of an Object: Simulation as Scientific *Factum*".

and Pitts, which von Neumann was quick and effective in pointing out—to the eventual resigned acceptance of McCulloch. Seemingly, von Neumann's suggestion to focus on the structure of very simple and primitive entities was confronted with the reality that they were not simple enough for mechanical re-creation. Once physical instantiation is out of the equation, cybernetics is left with little that would make it different and unique, let alone revolutionary.

CHAPTER 8

Cybernetic Tensions: Anatomy of a Collapse

THE IRRELEVANCY OF MATERIALITY FOR THE NATURE
OF A MACHINE: FROM COMPLEXITY TO DISEMBODIMENT

It has been indicated that several scientists in the UK were very interested in cybernetics right from the beginning, influencing their own careers and even forming a small, invitation only, regular discussion gathering.[1] Ross Ashby, producing his own insights squarely within cybernetic thinking at the start of the 1940s[2]—almost a decade before Wiener's catalyzing book[3]—seemingly performed a tune-up of the enterprise's theoretical core. Some years later, he recounted in his own book,[4] the intellectual adventure into the solidification of cybernetic science, providing valuable insight into his life's work—which was to some extent, as previously shown, apparently more concerned with cybernetic *dicta* than that of the original cyberneticians themselves.

A crucial point that Ashby found as fundamental to the whole enterprise was the notion of *control*. We have seen that Norbert Wiener was well aware

---

[1] The Ratio Club (Chapter 3, Section "The Ratio Club").

[2] Ashby 1940.

[3] Wiener 1948a.

[4] Ashby 1952a.

© The Author(s) 2017
A. Malapi-Nelson, *The Nature of the Machine and the Collapse of Cybernetics*, Palgrave Studies in the Future of Humanity and its Successors, DOI 10.1007/978-3-319-54517-2_8

of its importance, identifying James C. Maxwell's nineteenth-century mathematical treatment of governors[5] as the theoretical foundational document for cybernetics.[6] Wiener's famous book carried the noun "control" in its subtitle. However, Ashby identified that a previous, more fundamental notion, was the one that made possible control in the first place; this more primal notion was that of a *machine*. The strategic move that extended the notion of machine into realms heretofore untouched by it, is located here. This extension is ingeniously articulated by means of choosing geometry as a cybernetically suitable theoretical counterpart. Ashby described how this mathematical branch went from being ruled (and constrained) by physical instantiations, to actually containing them later on. This extension gained for geometry unprecedented *control...*

> There was a time when "geometry" meant such relationships as could be demonstrated on three-dimensional objects or in two-dimensional diagrams... In those days a form which was suggested by geometry but which could not be demonstrated in ordinary space was suspect or inacceptable. Ordinary space dominated geometry. Today the position is quite different... Today it is geometry that contains the terrestrial forms, and not vice versa, for the terrestrial forms are merely special cases in an all-embracing geometry. The gain achieved by geometry's development hardly needs to be pointed out. Geometry now acts as a framework on which all terrestrial forms can find their natural place, with the relations between the various forms readily appreciable. With this increased understanding goes a correspondingly increased power of control.[7]

Cybernetics follows this path: it has the same relation to an actual machine that geometry has to an actual, "terrestrial" object. Hence, cybernetics "takes as its subject-matter the domain of 'all possible machines,' and is only secondarily interested if informed that some of them have not yet been made, either by Man or by Nature".[8] In such way, Ashby managed to identify the crucial elements that make a machine what it is—its "essence"—thus furnishing its particular capacities. And the

[5] Maxwell 1868.
[6] Chapter 3, Section "Norbert Wiener's *Cybernetics*".
[7] Ashby 1956, p. 2.
[8] Ibid.

mechanization of entities entails the exertion of controlling power over them. For Ashby, such articulation was isomorphic with scientific explanation, in so far as one understands "explanation" in this context as isomorphic to providing an "explanatory *mechanism*". A mechanism entails a machine—and we managed to find it in nature. Modern science is on the right path.

Aware that this accomplishment is something remarkable in terms of extending powers of tractability to realms previously beyond mechanistic reach, Ashby claimed that it is "one of the substantial advances of the last decade [1950s] that we have at last identified the essentials of the 'machine in general'".[9] Here Ashby is in fact referring to himself, as he was correctly aware that "[m]any a book has borne the title 'Theory of Machines', but it usually contains information about mechanical things, about levers and cogs".[10] It is he who comes up first with a rigorous definition of what a machine is:

[A] machine is that [whose] internal state, and the state of its surroundings, defines uniquely the next state it will go to.[11]

This over-encompassing definition, the embodiment of a merge between a theory of machine behavior and negative entropy, gained unprecedented "control" for cybernetics by means of extending the reach of what can be understood as a full-blown machine. Ashby confided that in order to see clearly a machine for what it is, a behavior-based premise had to gain precedence, so as to dispel probable metaphysical or epistemological misunderstandings of what essentially constitutes a machine. And so he expressed that

[9] Ashby 1962, p. 260.

[10] Ashby 1956, p. 1. Also, see the introductory remarks to Chapter 6.

[11] Ashby 1962, p. 261. This might be, in the view of this writer, the only rigorous definition of a machine ever advanced. History of science does not seem to show any precedent. Alan Turing and Heinrich Hertz might have come the closest, but they still fell short of providing a full-blown definition (see Chapters 4, Section "A Machinal Understanding of an Algorithm and the Material Liberation of the Machine."; and 6, Section "William Ross Ashby's Nature-Machine Equalization", respectively).

Before the essentials could be seen, we had to realize that two factors had to be excluded as irrelevant. The first is "materiality"—the idea that a machine must be made of actual matter, or the hundred or so existing elements. This is wrong, for examples can readily be given . . . showing that what is essential is whether the system, of angels and ectoplasm if you please, behaves in a law-abiding and machine-like way.[12]

By musing about machines made up of "angels and ectoplasm", Ashby wanted to push the point across unambiguously: the materiality of a machine is entirely circumstantial. Not only has this observation closed the case regarding whether something purporting to be a mechanism is a machine—the answer being obviously in the positive. It also displaces the importance we have traditionally (both scientifically and philosophically) attributed to physicality—in the sense of materiality. In this context the physical sciences themselves might see their strong theoretical attachments to materiality seriously affected. The definition of the physically real would no longer entail material measurability. This seems to be a historic instance where an epistemological backhand move was surreptitiously passed. And it could have resulted in deep tensions within the cybernetic ontology—committed since the start to a physical grounding in a radically Viconian manner.[13]

It has been several times pointed out that reliance on physical instantiations of its own theoretical proposals was part and parcel of the cybernetic impetus. This was untenable after an anti-cybernetic "heresy" was committed. Once the "harm" was done, it became evident for both cases that the cybernetic mandate for physically grounding the theory complicated things to the point of no solution. Where Ashby had to concede even on his own terms an augmentation of complexity within the structure behind the homeostat's behavior (to be able to account for a more life-like comparison, thus giving birth to the doomed project of DAMS, where no order was to

---

[12] The quote continues: "Also to be excluded as irrelevant is any reference to energy, for any calculating machine shows that what matters is the regularity of the behavior—whether energy is gained or lost, or even created, is simply irrelevant" (Ashby 1962, p. 260).

[13] For Vico's philosophy see Chapter 9, Section "Viconian Constructability as a Criterion of Truth".

be attained), von Neumann had to plunge into the structure of complex entities (by-passing McCulloch and Pitts' networks, engaging into the construction of automata according to cybernetic dictum, only to then retreat into the non-physical due to the very complexity that made him turn away from the artificial neural networks in the first place).

Albeit with different technical motivations but with the same cybernetic intentions, both Ashby and von Neumann suffered the backlash of tinkering with actual structures—as opposed to manipulating models of their behavior. It is important to signalize that at this point both had already crossed the methodological line in the sand, entering into non-cybernetic territory. The investigation of the nitty-gritty stuff operating *under the hood* in the "teleologically behaving" mechanical systems is the task of traditional science practice, not of the new science of cybernetics, equipped which such an impressive novel theoretical edifice. Even if cybernetics did not have as an aim to transcend or replace physics, it did want to extend it—in its own eyes, to actually "defend" it.[14] The price to pay for this trespassing came from cybernetics itself.

When the cybernetic mandate for behaviorally articulating the studied object fell short, then entering the realm of structure to better understand the operating entity seemed to be the next, scientifically motivated, viable step. But this move clashed with the other cybernetic mandate of instantiating theories *in concreto*. More precisely, the mandate for buildability cancelled the possible import of going out of the cybernetic way into structural insights, given that this very need for physicality rendered the recreated structures as caricatures of the modeled object. After von Neumann's kinematic model failed he had to retreat into logical research. After Ashby's DAMS failed he ended up studying systems. Unexpectedly, the last cybernetic move was to find refuge in the realm of the disembodied, betraying a main column (arguably its strongest differentiator) of this "new science".[15]

---

[14] Chapter 2, Section "The Macy Conferences"—regarding the first Macy conference.

[15] Wiener 1948a, p. 28.

MECHANIZING MACHINE-UNFRIENDLY REALMS: SURREPTITIOUS METAPHYSICAL COMMITMENTS DIE HARD

Let us recall that the origin of cybernetics is situated in the context of research on anti-aircraft defense—the nature of the mechanically operated anti-aircraft gun, and the soldier's almost symbiotic relation with the operating machine, both processing in synchronicity the information available about the enemy target. The question "what kind of machine have we placed in the middle?"[16] was attacked by Wiener by means of a behavioral understanding of both the machines *and* the human in between them, under one and the same theoretical framework. This was made possible thanks to a more extended view of what a machine stands for, taking cues from Turing's suggestions regarding the mechanical essence of an algorithm—or the algorithmic essence of a machine.[17] Shannon's work on information theory anchored both realms even closer, both seamlessly crossed-through by information flow. Considering these black boxes as displaying active "aims" was the feature that distinguished cybernetic talk from the already entrenched behavioral psychology of the time.

The approach seemed promising, and to some extent, it flourished. Understanding adaptive—even intelligent—behavior from an exclusively external point of view (leaving inquiries into the nature of the working structures aside) produced striking results. There were the mice and the tortoises of Claude Shannon and Grey Walter, which attracted attention beyond cybernetic circles, occupying pages in mainstream magazines and booths in science festivals. There was McCulloch and Pitts' networks, which mathematized certain purported behaviors of the brain and the nervous system, serving as the platform for von Neumann's construction of the first stored-program computer. They seemed to have hit the sweet epistemological spot regarding the behavior of entities in nature, which would otherwise be riddled with vitalistic assumptions. Rigorous behavioral descriptions could be unambiguously measured. It seemed to be the right path chosen for understanding man, faithful to (and consistent with) the mechanistic tradition. After all, for the cyberneticians such tradition

---

[16] Edwards 1996, p. 197.

[17] Chapter 4, Section "A Machinal Understanding of an Algorithm and the Material Liberation of the Machine".

effectively was still the strongest one (if not the only one) in modern science, if we are to maintain that "to understand" a thing means "to provide the mechanism" for such thing.

The homeostat's creation was a necessary outcome of such a position— it was just a matter of time. Cybernetic tenets pointed in that direction right from the start. But cyberneticians might not have gone *that* extra mile. When Ashby entered the picture, it probably rang the death knell of a project whose radically materialistic ontology might not have shown itself with all its colors to the cyberneticists themselves. Mechanical mice and electric tortoises, wandering autonomously, avoiding obstacles, and "figuring out" their way out of a maze, were creatures that immediately were predicated of as being essentially the same as the ones that nature boasts—just simpler, but with possibilities of growth in complexity. The homeostat was another story. With this machine, the learning process of a kitten (who is first attracted to a flame, but after being burnt, begins to avoid it) does not qualitatively stand out from the situation of a plastic ball finding its way out of a moving box with a hole.[18] Under the hood, both situations where underpinned by random processes that would lock and capitalize on those outcomes that benefit the system—to the dismay of Julian Bigelow, who saw no virtue in that.[19]

Let us recall the damming examples of the young philosopher Richard Taylor when attacking Bigelow, Wiener, and Rosenblueth's 1943 article.[20] The behavior of a missile with a rope attached to its nose, being pulled by a submarine, would also count as behavior with an aim (along with that of a self-guided missile, the example given in the criticized article). Trivialization seemed to have been a looming outcome always present. In the case of the homeostat at Macy, trivialization seemed to have worked its way to those very foundations which lured brilliant minds into cybernetics. The cybernetic project was partly appealing because of the promise of being able to construct artifacts that would not only emulate, but eventually replace those constructed by nature—thanks to the discovered knowledge of a machine-basis common to both, the natural and the artificial. That

---

[18] Chapter 6, Section "The Backlash: Unforeseen Consequences of a Behavior-based Ontology".

[19] Ibid.

[20] Chapter 5, Section "Machines Can Be Teleological".

promise proved to be exciting and extremely attractive (probably more so in the context of a Cold War threatening the existence of natural life as we knew it). But then came a cyberneticist who faithful to cybernetic tenets *in extremis*, proclaimed that the very features that we understand as the ultimate pointers for life, namely self-organization and adaptation, are in no way particular to living entities[21]—in fact, they are a necessary condition of existence for material objects coping, such as crystals.[22] And randomly so at that. And he had a machine to show for it. Ashby, however, was not without a tradition of thought that heavily relied on experimentation and physical recreation.

Furthermore, this shift could have also signified the coming of an epistemic turn—still in the making[23]—inimical to a classical cybernetic epistemology: life as necessary order coming out of chaos, or negative entropy as the natural course of things coping in the world; both against what Wiener would defend.[24] Let us recall Ross Ashby's heated exchange at the 1942 Macy conference.[25] There, some in the audience were taken aback by Ashby's assertion that his homeostat displayed exactly the same learning and survival that a cat displays when it learns and survives. Ashby reminded them that "to learn", in cybernetic lingo, necessarily entailed, "objectively" speaking—without retorting to psychologisms or introspections—a *consequent change of behavior*. An "animal" can be observed to "learn" in order to "survive": In strictly physicalist terms that translates to a system adapting in order to reach and maintain equilibrium. To lose the capacity to adapt means to die. Seemingly, some cyberneticians (e.g., Julian Bigelow) were not too comfortable with the kind of conclusions that Ashby showed they were compelled to reach, if they were to still hold on to the basic tenets of cybernetics. To recall, Ashby indeed bridged the artificial inanimate with the organic living in no vague terms.[26]

---

[21] In words of Ashby, "a machine can be at the same time (a) strictly determinate in its actions, and (b) yet demonstrate a self-induced change of organization" (Ashby 1947b).

[22] Chapter 5, Section "Machines Can Be Teleological".

[23] Ibid.

[24] Chapter 5, Section "Machines Can Be Teleological".

[25] Chapter 6, Section "Un-cybernetic DAMS: From Behavior into Structure".

[26] Recalling the relevant quote by Ashby:

Ashby was relentlessly taking a machine-ontology based upon observable behaviors to its ultimate consequences, confident of its epistemological strength and methodological maneuverability. He showed this unmovable confidence throughout the years in his own theoretical development—if anything, the certainty grew stronger. Hence, at the Macy conference, he convincingly showed to the cyberneticians that if behavior, *tout court*, is what tells us what a thing is, then they have to concede that his presented machine is alive—or, at the very least, that living entities live and survive in the same way that his machine does. If one considers that a cat behaving differently after training can be said to have learned, one is compelled to accept that the machine has learned, given that it behaved differently after training.[27] This was in fact a crucial aspect of cybernetics.

Ashby was, however, somewhat encouraged by the criticism he received regarding the relative simplicity of the exposed entity (his machine), of which it could be said, through a exclusively behavioral lens, that it was indeed alive and learning—but not precisely "alive and kicking". More was needed. Mindful of the feedback received at the conference, Ashby allowed for higher layers of complexity in the structure of the machine to be an issue to be dealt with—so that it could be more justifiably comparable with the behavior of biological entities. In fact, he already had embarked upon the construction of DAMS.[28] However, this last one never managed to attain the kind of self-organizing adaptive behavior of the simpler homeostat. For a moment, while the homeostat showed its

> The moment we see that "adaptiveness" implies a circuit and that a circuit implies an equilibrium, we can see at once that this equilibrium must be of the stable type, for any unstable variable destroys itself. And it is precisely the main feature of adaptive behavior that it enables the animal to continue to exist... I must say unambiguously what I mean by "alive". I assume that if the organism is to stay alive, a comparatively small number of essential variables must be kept between physiologic limits. Each of these variables can be represented by a pointer in a dial... This, of course, is in no way peculiar to living organisms (Ashby 1953, p. 73).

[27] The "post hoc ergo propter hoc" fallacy was seemingly not flagged, possibly due to the context of the necessity of an exclusively mechanical explanation—which severely reduces the possibility of entertaining alternative explanations.

[28] Chapter 6, Section "Un-cybernetic DAMS: From Behavior into Structure".

glory, one could witness the argument for a predetermined state of nature where coping mechanisms inevitably reach self-organization—and eventually, life, likely to Wiener's chagrin—through trial and error, developing order out of chaos. Instead, the chaotic behavior of the newer more complex automaton never seemed to reach, let alone settle in, a coping pattern, despite 2 years of careful observation. Ashby, discouraged, abandoned the project and focused on the theoretical exploration of the notion of "cybernetics of cybernetics" (or second order cybernetics),[29] where the "observer becomes part of the system", engaging in projects without physical grounding and effectively going under the radar (both scientifically and publicly) until the end of his life.

A Machine's Isomorphism with Highly Complex Entities: The Intractability of an Absolutely Accurate Model

An exclusively behavioral-based understanding of a coping system was explanatorily adequate when the entity did not exceed a certain threshold of complexity. Once that threshold was crossed, then explanatory models would explode toward the practically unfeasible. During the Hixon Symposium,[30] McCulloch considered the possibility of having his networks utilized for mapping thought processes, accepting that the adequate implementation might now be in its infancy, but mostly due to an issue of availability of materials—and not due to the theory itself. Thus, we saw how McCulloch referred somewhat compassionately to von Neumann's physical instantiations of his networks—namely, the first stored-program computing machines.[31]

[29] Chapter 3, Section "The Decline".

[30] Jeffress 1951.

[31] To recall, McCulloch said that

> Neurons are cheap and plentiful. If it cost a million dollars to beget a man, one neuron would not cost a mill. They operate with comparatively little energy. The heat generatedraises the blood in passage about half a degree, and the flow is half a liter per minute, only a quarter of a kilogram calorie per minute for 10 10, that is, 10 billion neurons. Von Neumann would be happy to have their like for the same cost in his robots. His vacuum tubes can work a thousand times as fast as neurons, so he could

Von Neumann would more than concur with McCulloch's hyperbole. In fact, building the computing machines using McCulloch's (and Pitts') model helped him notice precisely such inadequacy, and in productive ways. Von Neumann saw in the very nature of the model—namely, one that maps network behaviors rather than structures—an inherent incapacity for capturing a very complex entity. It was a model surely amenable to be instantiated in a computing machine—as he efficiently showed—but thoroughly inadequate to map a brain. Von Neumann, thus, gave an image-rich (negative) verdict of McCulloch and Pitts' networks in regards to being a model suitable for understanding the nervous system.[32]

John von Neumann also expressed this dissatisfaction in the aforementioned infamous letter to Norbert Wiener[33] at two main levels. First, McCulloch and Pitts' networks seemed to be a model unfit for studying a highly complex object—knowledge likely acquired and/or reinforced with his own experience while building a computer based upon it (and which already occupied a physical space of considerable dimensions). Second, leaving the model aside, studying the nervous system *at all* seemed too far-fetched and even hubristic, given its extraordinary—perhaps the most extreme—complexity. Putting the two criticisms together, von Neumann fleshed out a "going back to the basics" proposal, that amounted to an epistemological turn also at two main levels.

match a human brain with 10 million tubes; but it would take Niagara Falls to supply the current and the Niagara River to carry away the heat (McCulloch 1951, p. 54).

[32] Recalling the relevant quote by von Neumann...

I think that it is quite likely that one may give a purely descriptive account of the outwardly visible functions of the central nervous system in a humanly possible time. This may be 10 or 20 years—which is long, but not prohibitively long. Then, on the basis of the results of McCulloch and Pitts, one could draw within plausible time limitations a fictitious "nervous network" that can carry out all these functions. I suspect, however, that it will turn out to be much larger than the one that we actually possess. It is possible that it will prove to be too large to fit into the physical universe. What then? Haven't we lost the true problem in the process? (Von Neumann 1951, p. 34).

[33] Von Neumann 2005.

First, to forego McCulloch and Pitts' networks of behavioral mapping. They have certainly proven their efficacy for other realms, such as automated computing processes, relieving humans from the corresponding tedious tasks. However, von Neumann encouraged shifting away from the observations of the behavior of the object in order to focus instead on its structure, given that after a certain point of complexity, the latter would be more amenable of tractability than the former. To recall, after a certain threshold of complexity, the gain of being able to rigorously describe a behavior simply disappears, since the complexity is so tremendous that the behavior description would be translated into a model of such dimensions that it would inherently escape any possibility of usability—hence effectively reversing the modeling roles. Attaining a full description of the elements constituting the studied entity would still be indeed difficult, but now it remains within the realm of feasibility—and that is a great procurement. However, von Neumann is introducing here a very un-cybernetic move, just as Ashby's stance regarding the necessary order coming out of chaos. It is worth taking another look at what von Neumann refers by "true" understanding of a complex object:

> I would, however, put on "true" understanding the most stringent interpretation possible:
> That is, understanding the organism in the exacting sense in which one may want to understand a detailed drawing of a machine, i.e. finding out where every individual nut and bolt is located, etc. It seems to me that this is not at all hopeless . . . I suppose (without having done it) that if one counted rigorously the number of "elements" in a locomotive, one might also wind up in the high ten thousands. Consequently this is a degree of complexity which is not necessarily beyond human endurance.[34]

The analogy here with a machine is not casual, and more so coming from an unapologetic cybernetician—which he surely was, despite his vocal disagreements. The immediate usefulness of the machine imagery here is the entailed amenability of being taken apart piece by piece—literally, one by one—as one could do, despite the daunting task, with an airplane or a ship. Cybernetics certainly aimed at understanding living organisms as machines; however, the extension of the notion of machine upon them was clearly

---

[34] Von Neumann 2005, p. 280.

based on the *behavior* of the former, and not upon the *structure* of the latter. As an outcome, for von Neumann, the still gigantic difficulty of the job would be now quantitative and not qualitative in nature—and that is where the "huge gain" is located. The complex entity would now be at least tractable—which is not the case with the unworkable number of variables resulting from the exact description of the behavior of a very complex object (like the brain). Such is the gain of turning into an "exacting sense"[35] of rigorously understanding something *qua machina*.

Second, since the ultimate structure of a complex entity is complex itself (the model approaching a one-to-one mapping upon the object), von Neumann recommended retargeting the cybernetic attention toward simpler organisms. However, we know that in the very letter of complaint against McCulloch and Pitts' networks (as a model for a brain), he already expressed his shocked amazement at the complexity of even structurally "simple" entities in the biological world.[36] Once we shift attention from a complex entity's behavior to its structure, the employed model risks entering into a one-to-one mapping with the objects' features, potentially bringing again an unsurmountable intractability. So he chose the simplest of them all: a virus. Despite the limbo area that it occupies in terms of fully counting as a living entity, a virus does display a feature essential to living things: reproductive capabilities. Precisely this capacity of self-replication is what von Neumann recognized and extrapolated as worthy to focus on and—in true cybernetic fashion—literally *build*.

Indeed, true to his cybernetic spirit—the same spirit that used the McCulloch and Pitts' networks for building the early computing machines—von Neumann embarked upon the task of recreating it *mechanically*. His "kinematic model" came to fruition, where an eventually constructed entity should be able to navigate through a host of floating resources, and use them to make a copy of itself. However, this

---

[35] Von Neumann 2005, p. 280.
[36] To recall, von Neumann candidly confessed that...

> After these devastatingly general and positive results [from the McCulloch and Pitts networks] one is therefore thrown back on microwork and cytology—where one might have remained in the first place...Yet, when we are in that field, the complexity of the subject is overawing (Von Neumann 2005, p. 278).

task was proven to be exceedingly more difficult than expected—arguably, until today. His resulting kinematic model, however, was fraught with matters pertaining to details inherent to material embodiment—to the way in which three-dimensional entities inhabit the world. Issues of solderability, separation, energy sources, friction, heat, material fatigue, and so on crippled the model right from the start. Being physical is hard, someone would say. This impasse pushed von Neumann to consider an alleged equivalent of this self-reproducing entity, but now entirely living in logical space. He subsequently found asylum in logical representations of lattices—a two-dimensional model that by its very nature sidesteps the problems entailed by the actual physicality of a self-reproducing automaton. In a manner that ultimately is not dissimilar from Ashby's late reaction, he dropped the physical instantiation of the model and opted for a theoretical project toward a new and enriched logic—leaving physical instantiations aside. After jumping out of the physical grounding into the abstract realm (calling for the development of a "new logic" more suitable for automata reproduction), the project failed to garner attention. Shortly after, von Neumann died. And the kinematic model remained in relative obscurity ever since.

Once cybernetics lost its unique methodological and ontological stance—namely, knowing by means of constructing and letting this construction lead to more knowledge—not much was left that was unique to cybernetics. Considering the renewed and sophisticated understanding of what a machine is, which constituted a cybernetic cornerstone at the time, it might come as somewhat expectable. After all, as we saw above, a machine does not have to be materially instantiated to be a machine, and so a physical embodiment of a machine is only tangentially important for the machine to be itself. However, physical instantiations were fundamental for the cybernetic enterprise from its inception. It would seem, thus, that a profound insight eventually led to its own demise. Under this perspective, one could see as the positive outcome of the cybernetic implosion a richer and deeper understanding of the nature of a machine.

However, there is an epilogue to this dramatic episode in the history of science.

CHAPTER 9

The Rise of Emerging Technologies

CYBERNETICS 2.0: THE NANO-BIO-INFO-COGNO CONVERGENCE

We saw how von Neumann's 1947 attempt to bring Max Delbrück into the cybernetic team was doomed to fail.[1] Delbrück went on to make substantial inroads in what later came to be known as "molecular biology".[2] This field was not entirely developed outside the cybernetic aura, however. A decade earlier, Claude Shannon's ex-boss, Warren Weaver (popularizer of Shannon's "Mathematical Theory of Communication")[3] was heading the Natural Sciences section of the Rockefeller Foundation. Weaver, who coined himself the term "molecular biology",[4] helped Delbrück coming to America in 1938, securing founding for him at Caltech. Around the time that Delbrück would be vowing to never come back to the Macy gatherings, Weaver was writing to Norbert Wiener about exploring a feasible application of Shannon's paper: Machine Translation—which drew substantial military interest from the

---

[1] Chapter 7, Section "The Letter to Wiener: Post Hoc, Ergo Propter Hoc?".

[2] Dronamraju 1999.

[3] Chapter 5, Section "Machines Can Be Teleological".

[4] Weaver 1970.

© The Author(s) 2017   217
A. Malapi-Nelson, *The Nature of the Machine and the Collapse of Cybernetics*, Palgrave Studies in the Future of Humanity and its Successors, DOI 10.1007/978-3-319-54517-2_9

government (an already familiar effect of cybernetics).[5] Thus, although Warren Weaver was not directly immersed into the cybernetic project, partly due to his role as science administrator (rather than practitioner), he was by proxy firmly related to the circle. In a 1933 Annual Report of his for the Rockefeller Foundation, one can readily see a cybernetic impetus *à-la-Weiner* regarding the future of humanity:

> Important questions are: Can we obtain enough knowledge of the physiology and psychobiology of sex so that man can bring this aspect of his life under rational control? Can we unravel the tangled problem of the endocrine glands and develop a therapy for the whole hideous range of mental and physical disorders which result from glandular disturbance? Can we develop so sound and extensive a genetics that we can hope to breed in the future superior men? Can we solve the mysteries of the various vitamins, so that we can, nurture a race sufficiently healthy and resistant? Can psychology be shaped into a tool effective for man's everyday use? In short, can we rationalize human behavior and create a new science of man?[6]

The attempt to finally rationalize the human condition would surely not look as an entirely new endeavor. The scientific and philosophical attempt at the "naturalization" of various aspects of what we refer as *human* arguably belongs to that drive. Cybernetics, as the science which looked forward to understanding animals as machines, was just a paradigmatic example of it—the culmination of an epistemological force coming to terms with its own fundamental tenets.[7] Part and parcel of this newer story is rooted in the collapse of cybernetics as a scientific proposal. This downfall begot brainchild scientific enterprises that would be reticent in acknowledging their cybernetic heritage.[8]

---

[5] Hutchins 2000.

[6] Rockefeller Foundation 1933, p. 199.

[7] Norbert Wiener's push for a mechanical common language between man and machine was relentless until the end—even connecting it later with issues of religion and Intelligent Design (Cf. Wiener 1964). For the connection between machinery and religion, see Chapter 9, Section "Baconian Subsumption of Nature to the "Mechanical Arts". For that of machinery and Intelligent Design, see Chapter 10, Section "Mechanizing Machine-unfriendly Realms: Blurring the Natural with the Artificial".

[8] Dupuy 2000, ch. 2.

Molecular biology evolved, receiving an important influx of scientists from basic science—physics. Delbrück proposed an understanding of genes based upon the information encoded in the morphology of crystals. Thus, advances in the technology of x-ray crystallography spurred Weaver for endorsing the migration of physicists to biology, having now at hand an important practical tool for testing the fallibility of novel bio-physical proposals.[9] Erwin Schrodinger was one of the leading physics figures that subsequently advocated for this naturalization of life—which served as a theoretical trampoline for James Watson and Francis Crick to device shortly after, in 1953, the D.N.A.'s double helix structure.[10]

By the 1960s, as global signatures of the Cold War, fear and mutual distrust between superpowers played important roles in "national identity building" scientific research.[11] These powerful drives would eventually result in placing a man on the face of the Moon, further develop our capacities for nuclear mutual annihilation and device the backbone for what we now know as the Internet. Among the projects that, in contrast, allegedly never came to fruition, another cybernetic unacknowledged off-spring stands out stridently: Artificial Intelligence.[12] Towards the end of the 1980s AI allegedly suffered its own first demise, abruptly losing funding, academic interest and overall traction.[13]

[9] For instance, the Irish crystallographer John Desmond Bernal (1901–1971) foresaw, more than a decade before John von Neumann's "epiphany", the possibilities opened up for mankind after a deeper understanding of structures at the smaller-than-microscopic level (Bernal 1929). More on this below.

[10] Dronamraju 1999.

[11] The same kind of "nation defining" research engaged by John von Neumann (Chapter 7, Section "John von Neumann's Appropriation of McCulloch and Pitts' networks").

[12] This writer remains agnostic regarding one of the most controversial, revisionist and emotionally charged topics in science and engineering of the last decades: The question whether current Artificial Intelligence, as it stands, lives up to the hype promised by its founders (see Chapter 3, Section "Traditional Explanations for the Collapse of Cybernetics"). For novel, arguably richer and perhaps more accomplishable versions of the notion of AI, see Chapter 9, Section "Transhumanism: An Age-old Grand Project for Humanity's Uplifting".

[13] This period, known as the "AI winter" was the direct outcome of frustration from founding sources (particularly the military), who judged that the AI promise

Also around this time, a major political event with planetary implications, whose consequences are still letting themselves be felt, took
place: the collapse of the Soviet Union. The subsequent rather abrupt
end of the Cold War entailed deep changes for science practice.
Without the looming threat of human annihilation, the alluded
"national identity building", gigantically expensive, scientific projects,
subsided—particularly in the physical sciences. Gradually, scientific and
academic interest, and thus, funding, shifted to the biological sciences.
The state, although not giving up control, substantially left more free-
way to market-driven forces. Arguably, due to this important turn in
availability of funding sources, a considerable drop in enrollment for
pursuing studies in physics and chemistry, particularly in the West, was
widely acknowledged toward the end of the millennium.[14] This general
realization has understandably triggered a multifaceted concern, if one
is to canonically think of science as a vehicle of progress and civilization.
It is in this context that we arrive at the following event.

Roughly two months after the September 11, 2001 attacks in US soil,
the National Science Foundation called up a meeting in Arlington,
Virginia, that gathered 200 scientists and technologists from widely
different disciplines.[15] The aim was to produce a unified report on the
future of scientific practice for the next decades. The approximately 500
pages-long resulting document was entitled *Converging Technologies*

simply failed to deliver. This was epitomized in the shutting down of the Strategic
Computing Initiative, the biggest governmental source of funding for AI—after
having invested $100 billion during the span of a decade—in 1988. For the
historical struggle of AI research, see McCorduck 2004, Afterword.

[14] Fuller 2011, ch. 3.

[15] There was a pre-meeting that took place the preceding Summer, so it would be
inaccurate to assume that the meeting occurred as a response to the attack.
However, given the military content of some of the proposals, which arguably
inspired the Institute for Soldier Nanotechnologies currently housed—fittingly—
at MIT, it is reasonable to suggest that it might have affected the tune of
subsequent editions of the discussion drafts, which took several weeks to finalize.
There is an entire section on "National Security", including presentations on the
possibilities for cognitively and bodily enhancing the infantry, and on how these
ameliorations would fit into future counter-terrorist operations. See National
Science Foundation 2003, pp. 327–362.

for Improving Human Performance: Nanotechnology, Biotechnology, Information Technology and Cognitive Science.[16] This report set the formal beginning for what was later dubbed the "Nano-Bio-Info-Cogno" (NBIC)[17] Convergence—a coordinated effort to synchronize current sciences and technologies toward an intended qualitative leap in human history. The hopes portrayed in that report were high, not dissimilar to those present in the early phases of cybernetics:

> We stand at the threshold of a *new renaissance* in science and technology, based on a comprehensive understanding of the structure and behavior of matter from the nanoscale up to the most complex system yet discovered, the human brain. Unification of science based on unity in nature and its holistic investigation will lead to technological convergence and a more efficient societal structure for reaching human goals... Convergence of diverse technologies is based on material unity at the nanoscale and on technology integration from that scale... Developments in systems approaches, mathematics, and computation in conjunction with NBIC allow us *for the first time* to understand the natural world, human society, and scientific research as closely coupled complex, hierarchical systems. At this moment in the evolution of technical achievement, improvement of human performance through integration of technologies becomes possible. This is a broad, cross-cutting, emerging and timely opportunity of interest to individuals, society, and humanity in the long term.[18]

It may not be far-fetched to consider the above paragraph an outright "manifesto". The allusion to a "new renaissance" is not rhetorical. The report identifies the historic success of the Renaissance as being based

---

[16] National Science Foundation 2003.

[17] "The phrase 'convergent technologies' refers to the synergistic combination of four major 'NBIC' (nano-bio-info-cogno) provinces of science and technology, each of which is currently progressing at a rapid rate: (a) nanoscience and nanotechnology; (b) biotechnology and biomedicine, including genetic engineering; (c) information technology, including advanced computing and communications; and (d) cognitive science, including cognitive neuroscience." National Science Foundation 2003, pp. 1–2.

[18] National Science Foundation 2003, p. 2. Italics added.

upon its own transdisciplinary capacity[19]: A Renaissance man (e.g., Leonardo Da Vinci) was a true polymath of implementation—multi-skilled at theoretical and practical levels. In contradistinction with this example, a serious obstacle for current scientific progress is the extreme specialization (without cross-linked communication) of scientific practice these days. Against this suboptimal situation, it is advanced as a premise that nature has itself some ontological unity, and that such unity can have its scientific counterpart at the nanoscale level of reality—1 billionth of a meter. Such realm, even if extremely small, still operates under Newtonian physics, so issues of quantum indeterminacy do not obtain. From that fundamental level up, the other three areas (information, biology, and cognition) should structurally fall in place, thus accomplishing "for the first time" this unified view of reality. The summarizing motto for this über-holistic approach would be:

> If the Cognitive Scientists can think it,
> the Nano people can build it,
> the Bio people can implement it, and
> the IT people can monitor and control it.[20]

To be sure, the motivation behind the mega-project was openly and unapologetically one of usefulness for scientific progress—on the antipodes of a Kuhnian isolationist stance of justified puzzle-solving independent of the social aspects underpinning scientific practice. The need for progress vis-à-vis a tangible need for human welfare (e.g., the cure of cancer) is more in tune with what Jürgen Habermas calls the finalizationist role of the state regarding the guidance of science toward the betterment of the human condition.[21] After the unity that will gradually be achieved as the outcome of scientific progress, there should be a functional and social role fulfilled. These changes shall be ultimately positive, profound and unsettling—probably less disruptive if exerted gradually. Some outcomes following these improvements will entail

---

[19] Recall that this was an important aspect of cybernetics' *raison d'être* (Chapter 2, Section "The AA-Predictor").

[20] National Science Foundation 2003, p. 13.

[21] Fuller 2011, ch. 3.

enhancing individual sensory and cognitive capabilities,... highly effective communication techniques including brain-to-brain interaction, perfecting human-machine interfaces including neuromorphic engineering for industrial and personal use, enhancing human capabilities for defense purposes, reaching sustainable development using NBIC tools, and ameliorating the physical and cognitive decline that is common to the aging mind.[22]

Remarkably, the allusion to "improvement of human performance" has a clear reference throughout the document to the notion of "enhancement", which appears dozens of times: not just augmentation of the "performance" or skills, but enhancement of the person itself. What is sought after is not a therapeutic restitution to normalcy, but a transcendence into an improved and superior form of being, such that...

the definition of human enhancement may entail providing people with advanced capabilities of speed, language, skill, or strength beyond what humans can perform today. Just as plastic surgery and pharmacology have given new choices to human beings today, enhancement treatments will no doubt shape tomorrow.[23]

The report, supported by the National Science Foundation, was co-edited by Mihail Roco and William Bainbridge—who conform a team reminiscent of both Weaver's role in the Rockefeller Foundation and Wiener's own at the Jociah Macy Jr. Foundation. Roco is an engineer and science bureaucrat; Bainbridge is a sociologist of religion and advocate of transhumanism.[24] This distribution of roles becomes all the more interesting when compared to that of its European counterparts, whose report was largely a reaction against the American initiative. The European report was commissioned to Alfred Normann,[25] a philosopher of science, and later to Steve Fuller,[26] a sociologist of science. This contrast, an engineer and religious studies scholar on the American side, and a philosopher and sociologist on the European one, might to

---

[22] National Science Foundation 2003, p. 1.

[23] National Science Foundation 2003, p. 92.

[24] More on this in the next section.

[25] European Commission 2004a.

[26] European Commission 2008.

some extent advance the nature of their markedly different approaches regarding the converging technologies agenda.

Roughly, the European Union's stance, although acknowledging the usefulness (even necessity[27]) of this upcoming convergence, is nonetheless more cautious in its embracement—against the more risk-friendly approach coming from the USA. The European Commission's seal of difference was manifested in the emphasis on the social outcomes, rather than on the individual enhancement, that one is to expect from the implementation of the converging technologies agenda. Not too keen on altering what we traditionally understand as *human* (particularly regarding the integrity of our physical bodies and our naturally endowed cognitive abilities), the positive enhancement should rather contribute to the improvement of the overall welfare of a nation.[28] This contrasting approach has been subject of much debate during the last decade, and the positions seemed to further clarify their differences instead of coming to a common agreement.[29]

The American report triggered a tangible leap in nanoscale science research, evinced by the prompt establishing of the US National Nanotechnology Initiative (NNI)—proposed by Roco himself, now Chair of the committee on Nanoscale Science, Engineering and Technology of the US National Science and Technology Council. This long-term project was signed by President Bill Clinton and

---

[27] One practical way in which a substantial improvement of our physical condition can be pitched to (particularly Welfare oriented) nation states pivots around the possibility of staying healthier for longer time, thus pushing back the receipt of retirement pension. The framework of retirement at 65 years of age came about when people used to live in average until their early 60s. People can now live 30 to 40 years after that same threshold, and soon half a century beyond that point. The burden to be put on economic systems (particularly in industrial nations without booming natality rates) will be enormous—perhaps unbearable. In face of this, there is a clear need in making people healthier so that they can willingly work more years before retiring (Fuller 2013a, pp. 115–124).

[28] Canada, through a report elaborated by its Ministry of Defense's Research and Development, sided with the American position (Defence Research and Development Canada 2003).

[29] This dualism is now instantiated in competing precautionary versus "proaction-ary" stances in science policy. More on this in the next section.

followed up by President George Bush, who increased funding. President Barack Obama approved US$ 1.5 billion in funding in 2015.[30] There is widespread consensus that the consequences of the breakthroughs at the nanoscale will have at least the disruptive effects that the information revolution did with the Internet.[31] Indeed, for the NNI, "the ability to understand and control matter at the nanoscale leads to a revolution in technology and industry that benefits society".[32] Not unlike the way in which the Cold War context of cybernetics set the pace of society during the 60s and 70s, the "nano" perspective shall shape not only science policy but culture itself, after which the social implications will let themselves be felt.[33]

[30] National Nanotechnology Initiative 2014a.

[31] Just as cybernetics and AI after it, nanotechnology had a brief moment of hype followed by a crash. The grandiosity of some expectations, coupled with extra-optimistic predictions of the future (recall the experience of cybernetics on this) lead some scientists to adopt an extra cautious view against the promises of nanotechnology. For instance, Eric Drexler foresees the construction of "molecular assemblers", namely, machines that will literally manipulate physical reality, transforming it at will by means of reaccommodating atoms—having as a consequence, among other things, ending the world economy as we know it. While some scientists believe that nothing in principle preclude their future construction, some others (e.g. Richard Smalley and George Whitesides) severely criticised its feasibility—adducing problematic effects that quantum mechanics still have on non-subatomic realms. Others would argue that nanoscience is nothing other than a nominalist move that rebrands good old chemistry under a new aura of cool edginess, in order to revert the mentioned avoidance of science careers happening in the West. Lastly, the aforementioned "War on Terror" demanded more readily available outcomes for defense, instead of "exotic" research and development. However, once the Global Financial Crisis of 2007–2009, which made funding scarce, begun to subside, sources for nanoscience research gradually returned and even found more applications—for example, material resistance for anti-terrorism defense. For Drexler's views on nanotechnology, including the criticism by Richard Smiley, see Drexler 1986. For an account of the said "hype" before the Global Financial Crisis, see Berube 2005.

[32] National Nanotechnology Initiative 2014b.

[33] For instance, Weiner's understanding of a soldier as a certain kind of machine—to be fit into a bigger machine—would hardly trigger today reactions of uncanniness vis-à-vis contemporary "super-soldier" research. Also, it is difficult not to

Expanding on this calculated awareness, the current situation is no less intriguing—although, to some extent, expected. Countries belonging to the increasingly independent[34] BRIC economic block (Brazil, Russia, India, and China) have each vowed to relentlessly invest into a future where nanoscale science will conform their paradigm of research. The migration of scientists toward those areas is well under way. Each country has its own tune on how the ensuing technological "convergence" will take place.[35] The USA and the EU claim that the most "quality" of research still stems from themselves, whereas BRIC countries (particularly China) already produce the most "quantity", in terms of published articles.[36] However, even this seems to be slowly changing, as a more mobile immigration is occupying science departments in host countries. These shifts in global knowledge influx will surely entail interesting (or disrupting, depending on who regards them) outcomes regarding the traditional web of strings exerting power around the planet.[37]

Thus, a driving force behind this unified aim of research seems to be very much present. And with it, an epistemological impetus, a "metaphysical research programme",[38] is very much alive. Not unlike the inherited Cold War scientific era that partly stood as a legacy of cybernetics, converging

perceive a substantial difference between the amazement, even fear, that people displayed at Shannon's tortoises, with the seamless familiarity that children have now with robotic entities. For the current mechanizations of next-generation soldiers—including its ethical implications—see Galliott and Lotz 2016. For a contemporary study of children's familiarity with robots, see Latitude 2012.

[34] Wallerstein 2003.

[35] Fuller 2011, ch. 3.

[36] As of 2015, China has the lead in terms of sheer amount of nanotechnology publications, while the USA has the most amount of "indexed" (in established scientific journals) publications (StatNano 2016, pp. 4–17). Germany, Japan, and the USA consistently occupy the first three positions in terms of nanotechnology patents approved (StatNano 2016, pp. 28–37). Also see European European Commission 2008, pp. 10–11.

[37] National Intelligence Council 2012, pp. 15–19.

[38] This is Karl Popper's term, controversially used when he referred to Darwinian evolution (Popper 2005, pp. 194–210; Sonleitner 1986). The sense in which it is employed here is the designation or identification of underlying metaphysical commitments behind scientific theories and practices (Dupuy 2004).

technologies will likely set the pace for the political management and administration of near future scientific practice—partly, but substantially, due to our more cybernetic-friendly human ecosystem, where canonically sanctioned traditional metaphysical values seem to have loosen their grip. In particular, some of the (mostly metaphysical) problems that both Ashby and von Neumann had to deal with at their times, today would seem to either vanish or at least get dramatically reduced, as we will see later.

However, let us first address the "driving force" underpinning the technological convergence.

TRANSHUMANISM: AN AGE-OLD GRAND PROJECT FOR HUMANITY'S UPLIFTING

We know by now that Norbert Wiener became increasingly interested in the possibilities offered by cybernetics in what concerns the amelioration of the life conditions for soldiers who suffered amputations. He worked on the prosthetics that served as extensions of their lost limbs—foreseeing a future where these extensions would actually be an improvement over the natural ones.[39] Such was one of the very useful outcomes of successfully identifying man with machine: Once man could be tinkered with using the same scientific laws that we apply to the mechanizable, the enhancement of man follows. In this context, we also saw how Wiener confided with his friend the geneticist John Burdon Sanderson Haldane his thoughts on the shortcomings of behaviorism regarding machinery.[40] Indeed, Wiener's friendship with the geneticist points to an interesting connection between Wiener's ideas and early transhumanist concerns.

Today, transhumanism could be understood as a global movement that aims at the amelioration of the human species by means of a deep alteration of its condition at the bodily and cognitive levels. The profound modification of our minds and bodies would lead human existence toward a society freed from the usual mental and physical sufferings that one expects from ignorance, disease, and death. In fact, some versions of transhumanism foresee leaving humanity as we know it altogether, in order to enter a post-human condition.[41] The movement is constituted by thinkers from almost all realms

---

[39] Mann 1997.

[40] Chapter 2, Section "The AA-Predictor".

[41] More on this below.

of life, including philosophy and the humanities, social sciences, basic and applied science, engineering, theology, and the arts. Although the term in its current iteration has been used for three quarters of a century, the fundamental tenets of the movement can clearly be found in major themes reoccurring throughout the history of Western philosophy. In fact, cybernetics itself can be understood as the boldest attempt so far at instantiating some transhumanist basic premises—as we will see.

Wiener, a Slavic Jew himself, confided with Haldane his thoughts on why Jewish people tend to show above average intellectual prowess. For millennia, while bright people among Christians were customarily hand-picked for receiving the Holy Orders[42]—hence cancelling the possibility of progeny due to the disciplinary mandate of celibacy present in the Latin rite of the Catholic Church[43]—Rabbis, who were also chosen among the intellectually gifted in the community, had the tendency to procreate even more than the average Jew. This situation established a form of artificial selection throughout the centuries that resulted in the famed Jewish intellectual power. Or so Weiner thought—and Haldane agreed.

Indeed, Haldane is credited with being the first thinker to clearly refer to future scientifically plausible human biological alterations that would benefit our existence on a permanent basis. In particular, he foresaw the possibilities of genetic enhancements that would improve the mental and physical capacities of both individuals and societies as a whole, by means of an artificial (as opposed to *only* "natural")[44] selection. Haldane's ideas on

---

[42] "Holy Orders is the sacrament through which the mission entrusted by Christ to his apostles continues to be exercised in the Church until the end of time: thus it is the sacrament of apostolic ministry. It includes three degrees: episcopate, presby-terate, and diaconate." (Catholic Church 1999, § 1536)

[43] "All the ordained ministers of the Latin Church, with the exception of perma-nent deacons, are normally chosen from among men of faith who live a celibate life and who intend to remain celibate 'for the sake of the kingdom of heaven.' Called to consecrate themselves with undivided heart to the Lord and to 'the affairs of the Lord,' they give themselves entirely to God and to men. Celibacy is a sign of this new life to the service of which the Church's minister is consecrated; accepted with a joyous heart celibacy radiantly proclaims the Reign of God." (Catholic Church 1999, § 1579)

[44] Nature would still do its part, but "guided" by human intervention. After all, Haldane is credited with helping to unify Darwinian evolution with Mendelian

the topic were vented in a series of lectures that later got compiled into a 1923 book entitled *Daedalus; or, Science and the Future*.[45]

Haldane's book served as inspiration for John Desmond Bernal, the crystallographer mentioned in the previous section, to write his own. The title of Bernal's 1929 book, *The World, the Flesh & the Devil: An Enquiry into the Future of the Three Enemies of the Rational Soul*, is a play-on-words after the doctrine of the "three enemies of the soul" (*mundus, caro et diabulus*) present in Catholic moral theology.[46] This time, straightforward transhumanist themes, as they recur until our days, were examined: The possibility of altering our bodies and modifying our cognitive abilities so that they better fit into space exploration ... The construction of spatial artificial habitats where humans can flourish free from the conceptual boundaries idiosyncratic to a retrograde planet ...[47] The social consequences to emerge after some individuals become 'enhanced' ... These were some of the farsighted preoccupations that Bernal entertained, and although the word 'transhumanism' is still nowhere to be found, his implementational angle to Haldane's proposals makes him also a sort of intellectual father of the movement.

Haldane's book also served as a basis for Aldous Huxley's famed 1932 *Brave New World*—which depicts a twenty-sixth-century London sharply stratified in classes each based upon genetic makeups controlled by the state.

genetics. Haldane's contribution to the "new evolutionary synthesis" took place via a series of talks compiled into a book entitled *The Causes of Evolution* (1932) as well as ten essays that conformed the series *A Mathematical Theory of Natural and Artificial Selection* (1924–1934).

[45] Haldane 1923.

[46] Aquinas 1273, p. III, q. 41, a. 3. Bernal, once a devout cradle Catholic, gradually left his faith while being enticed by communism in general and the Soviet Union in particular. Later in life, Marxism also fell prey to his scientific scepticism—although he maintained an everlasting sympathy for socialism. See Brown 2005, chs. 2 & 4.

[47] The spirit of looking up to the skies, instead of down to Earth, gradually became, beyond a reference for space colonization, rather the visual anchoring of its social philosophy and ethics. The Iranian futurist Fereidoun Esfandiary (1930–2000) articulated this view in his book *UpWingers: A Futurist Manifesto* (1973), advancing a sort of "90 degree turn" on the political axis away from the traditional left/right divide. In the same spirit, lately Steve Fuller has been further developing it (Fuller 2013b). Also, see the "proactionary" notion below.

This rather dark take on a human future underpinned by transhumanist ideals is contrasted by the views of his brother, Julian Huxley, the first thinker canonically credited (wrongly[48]) with coining the term "transhumanism".

If the specter of eugenics seemed to rear its head with Haldane and Bernal, Julian Huxley openly advocated for the amelioration of the human species. In fact, his attempt at replacing the term "race" for "ethnic group" while presiding the United Nations Educational, Scientific and Cultural Organization (UNESCO)[49] fell into this agenda. Huxley wanted the improvement of the "human race" as a whole, as opposed to the version cooked up by Nazism, which identified one race as superior to others—and whose disastrous consequences gave eugenics a bad reputation up to these days. Thus Julian Huxley, keenly aware of the supremacist danger, presented the grand ideal for the betterment of our species, baptizing it in a 1951[50] lecture:

> Such a broad philosophy might perhaps be called, not Humanism, because that has certain unsatisfactory connotations, but Transhumanism. It is the idea of humanity attempting to overcome its limitations and to arrive at fuller fruition.[51]

Although the above is a famous passage where the term "transhumanism" is regarded as emerging "for the first time", the term was actually coined and introduced a decade earlier by the Canadian philosopher William Douw Lighthall. In 1940, Lighthall referred to the Apostle St. Paul's experience of a type of reconfiguring knowledge that transcends physicality, by-passing sensoria, as "St. Paul's Transhumanism".[52]

[48] Harrison and Wolyniak 2015.

[49] UNESCO 1950, p. 6.

[50] Some authors (e.g. More 2013a, p. 8; Tirosh-Samuelson 2011, p. 20) regard a 1957 paraphrase of the quoted paragraph as the first instance where Huxley used the term. Still others locate it as early as 1927. Both accounts are wrong, according to Harrison and Wolyniak (2015, pp. 465–466). Also, Max More (ibid.) claims that Thomas Stearns Eliot used the term "transhumanized" in his play *The Cocktail Party* in 1935. Elliot wrote that play, at the earliest, in 1948 (Institute for Advanced Study 2007, p. 6).

[51] Huxley 1951, p. 139.

[52] Lighthall 1940, p. 139. Lighthall was referring to the Biblical verse:

Furthermore, not unrelated to the Pauline extra sensorial experience, already in 1814 we can find an English version of Dante Alighieri's *Divine Comedy* where Rev. Francis Henry Cary[53] chose to translate as an adjective the novel Italian verb *trasumanar*. Dante coined this word for his manuscript, referring to the act of transcending humanity in its transformative participation with the Divine (action that, in Dante's mind, the Apostle experienced better than most)[54]:

> What no eye has seen, nor ear heard, nor the heart of man conceived, what God has prepared for those who love him (I Corinthians 2:9, RSVCE).

[53] Harrison and Wolyniak, apparently in error, name the Victorian translator as "Francis Henry Carey", not "Cary", in the body of their article—and as "William Henry Carey" in its footnote (Harrison and Wolyniak 2015, p. 467). The name of the famed British writer and translator of Dante was (Reverend) Francis Henry Cary (1772–1844). Cary's place in history is in fact worthy of attention. Dante's *Comedy* was not translated into English, due to an Anglo-Protestant dislike of Continental Catholicism, for almost five centuries! After all, Dante not only refers to St. Thomas Aquinas throughout the work with utmost reverence, but he uses hundreds of citations from both the *Summa Theologiae* and *Summa Contra Gentilies* (Mahfood 2015). Indeed, Dante's philosophy is widely regarded as Thomistic through and through (Lafferty 1911). And Thomas Aquinas was the very saint and Doctor of the Church whose work was used to define the Catholic position against the Protestant revolt, placing his *Summa Theologiae* on an altar, next to Scripture (the only book ever granted such honour) during the deliberations of the Council of Trent (Leo XIII 1879, 22). The fact that Cary was an Anglican minister is all the more remarkable, since he pursued the second ever full translation of Dante's *Comedy* into English (The first one took place around a decade earlier).

[54] Harrison and Wolyniak believe that Lighthall committed a mistake in alluding to I Corinthians 2:9 as the Pauline epistle to which Dante was referring. Instead, Dante was ostensibly referring to II Corinthians 12:4: "...and he heard things that cannot be told, which man may not utter" (RSVCE). They are not alone in such opinion (Langdon 1918, vol. 3, p. xxvi). Harrison & Wolyniak claim that Lighthall might have gotten confused between "no eye has seen, nor ear heard" and "things that cannot be told, which man may not utter", switching the verses by mistake. In all fairness to Lighthall, although one can claim that the passage refers to the spiritual knowledge hidden to "the rulers of this age" to which its immediately previous verse refers ("None of the rulers of this age understood this"—I Corinthians 2:8, RSVCE), Lighthall could also be referring, if more broadly understood, to the

Words may not tell of that transhuman change:
 And therefore let the example serve, though weak,
 For those whom grace hath better proof in store.[55]

Be that as it may, it would seem that both the term and the notion of transhumanism have been with us already for a while. One can discern throughout its various instances in history a concern with the uplifting of humanity. This preoccupation is apparently present from its inception, whenever that was, and it certainly did not leave theology behind.[56] Precisely this "prime directive" concerning the improvement of the human condition determines the connection (or lack of it) with the "posthumanism" advanced above. Some thinkers see a future where humanity is so evolved, biologically and cognitively, that the case where

non-sensorial, mystical, and reshaping knowledge that St. Paul was acquiring from his conversion in Damascus on. The Apostle emphatically refers to the creation of a "new man" over an "old man" (Romans 6:6; Ephesians 2:15; Ephesians 4:22–24; and Colossians 3:9–11) which could fit with Lighthall's notion of "St. Paul's Transhumanism". Indeed, the Catholic Academy of France has recently explored in a conference the link between such *metanoia* and transhumanism *per se* (Académie Catholique de France 2015). After all, mystical knowledge has on occasion triggered a philosophy—e.g., among Franciscans—that has fostered trans-humanist outlooks, as we will see in the next section.

[55] Trasumanar significar per verba
 non si poria; però l'essemplo basti
 a cui esperïenza grazia serba (Canto I, vv. 70–72).

[56] The Medieval times provided fertile ground, due to its emphasis on mysticism and transcendence, for transhumanist themes—an instance of which to be shown in the next section. A further witness to that connection with the mystical, beyond its Western Medieval sources, is both the influence from Russian Cosmism—a worldview that stems from Christian Orthodox mysticism (Young 2012)—and the widely acknowledged influence from the Jesuit Paul Teilhard de Chardin (Steinhart 2008; Delio 2012). There are those who rather find Renaissance humanism as a direct ancestor of transhumanism. Still there are others who find the Nietzschean "Übermensch" as a proto-transhumanist instantiation (Sorgner 2009). It would seem that transhumanist longings emerged scattered throughout the very evolution of Western civilization. For an account of transhumanism attentive of its religious roots, see Cole-Turner 2011 and Mercer 2014. For accounts of transhumanism that put more emphasis on its secular sources, see Bostrom 2005 and More 2013a.

one should no longer be called "human" may obtain (e.g., an individual who is mechanically or genetically grafted in more than 50% of its being). In this case, posthumanism would be the end game of transhumanism.[57]

In contrast, others see a subtler, arguably more profound relation between transhumanism and posthumanism. And this one is a relation of hostility. In this latter case, posthumanism would stand for an "evolutionary-friendly" view where humans are bound to disappear from the face of the planet, as per the laws of natural selection. Our demise would be indeed good news for both the Earth—after our unbridled pollution, exploitation of its resources and the disastrous effect of both on its ecosystem—and the species that would replace us as masters of this great habitat. This position is in line with Darwin's famous regard for the human species as *not* superior to any other one—and thus, subject to the inner workings of an ultimately inscrutable "Mother Nature", whatever those turn out to be.

In contrast, the thinkers that pitch transhumanism against posthumanism recoil at the above prospect and unambiguously assert a human supremacy of sorts over and above other species.[58] Indeed, "Nature" would not be "Mother", let alone "Master", but a *problem* to deal with, in order to guarantee the flourishing of the human species.[59] Thus, the transhumanist core of ecological concerns would be unapologetically anthropocentric. In fact, the survival of the human species vis-à-vis imminent ecological catastrophes, or unhinged human-unfriendly technological advancement,[60]

---

[57] For a collection of essays friendly to this view, see Ranisch and Sorgner 2014. There are those within this view who see "postgenderism" (the biological collapse of sexual differentiation) as part of a post-human future (Dvorsky and Hughes 2008).

[58] Arguably the most articulate defense of this dichotomy is found throughout Steve Fuller's transhumanist trilogy: Fuller 2011, Fuller 2013a, and Fuller and Lipinska 2014.

[59] For the deep correlation between Intelligent Design in the broad sense, and the preoccupation with putting humanity above nature (against a "human-unfriendly" radical Darwinism), see Fuller 2007, 2008.

[60] The popular notion of the technological singularity has to do precisely with this dystopian situation. In its general conceptualization, the singularity refers to the moment when technological advancement is so steep that humans would have to become a sort man-machine hybrid just to be able to survive quotidian life. Friendly to a cybernetic view of man, in this scenario Artificial Intelligence would occur within the human skull and its possible cognitive extensions. Perhaps the best-known articulation

is a major motivation for research on transhumanist technologies. Equipped with the disruptive tools of converging technologies, the approach would be one of "participatory" or "directed" evolution,[61] so that humans, after our steering of nature, could ultimately survive and flourish—most likely in deeply altered bodies. The technologies of interest for a transhumanist agenda are indeed the ones addressed in the NBIC reports mentioned earlier: nanotechnology, biotechnology, information technology, and cognitive science.[62]

of the singularity is currently advanced by the Director of Engineering at Google, Raymond Kurzweil (Kurzweil 2005). For the connection between Kurzweil and a cybernetic view of future humanity, see Fuller 2013a, ch. 1.

[61] The idea of anthropocentrically "steering" evolution is lately underpinned in part by research into "horizontal gene transfer"—the ability to share genetic code without the use of reproduction. Carl Woese (1928–2012), the discoverer of the third kingdom in the Tree of Life—Archea—maintained that there was an early evolutionary era when genetic code was shared between unicellular organisms in such manner. Only much later primitive organisms stopped sharing their genetic material "horizontally", making use of reproduction instead (vertical gene transfer). This made evolution a much slower and less efficient operation—Darwinian evolution properly referring to this latter period. In fact, the identity-relation currently existing between "Darwin" and "evolution" is for Woese an unfortunate situation:

> [W]e regard as rather regrettable the conventional concatenation of Darwin's name with evolution, because there are other modalities that must be entertained and which we regard as mandatory during the course of evolutionary time (Woese 2007, p. 369).

The prospect of a more efficient, "non-Darwinian" evolutionary process, already utilized by nature, made Woese suggest the extension of biology into a broader and richer discipline: "synthetic biology". This proposal is later taken up by Freeman Dyson, who sees genetic engineering as a type of horizontal gene transfer, indeed ultimately aiming at returning nature to the path from which it went astray. See Woese 2004, Woese 2007, and Dyson 2007.

[62] Jean Pierre Dupuy finds it remarkable that of the three areas of knowledge, three are *technologies* and one is a *science*—the one pertaining to the mind. He sees thus cognitive science as the ordering beacon for the convergent technologies to fall in place correctly (Dupuy 2007, pp. 250–251). However, in what concerns these technologies, the line between what is scientific and what is technological is getting blurred (in part due to the fundamental importance of the tools and methodologies used in contemporary scientific practice, about which more will be said in the next chapter). One can

Areas of special interest stemming from these are genetic engineering, synthetic biology, simulation, man-machine interfaces, robotics, and cryonics.[63] Expectably, the political sentiment behind the bold proposal for the widespread use of disruptive technologies with the purpose of profoundly altering the human condition—entailing the right to defend our morphological freedom[64]—cannot be a conservative one. Given that conservatism usually opposes attempts at a "social engineering" of any sort, it is not a liberal one either. Transhumanism is fostering a novel political agenda with a proactionary[65] signature, as opposed to a precautionary stance. This is the social and political sentiment seemingly required to provide fertile

gradually see these areas of research being addressed as nanoscience, bioscience, information science, and cognitive technologies, respectively.

[63] Cryonics, the deep freezing of a recently deceased body, in order to reanimate it in a future where medicine is more advanced, is a technology of interest for many transhumanists. Furthermore, some foresee the possibility of radical life extension by means of either halting, or even reversing, the aging process (De Grey 2004). This emphasis on preserving or enhancing the body to conquer decease and death is contrasted, within transhumanism, with a more futuristic view, wishful of eventually abandoning our carbon-based embodiments (by means of a future mind uploading of some sort) in order to have an existence *in silico*—in a supercomputer, in "the cloud", and so on (Kurzweil 2005, ch. 4). It is worth noting that many transhumanists, primarily from a religious extraction, dismiss this aspect of the project as dystopian, fantastic or simply undesirable, considering it as non-essential for the grand-project of the amelioration of the human race (See Waters 2011; Daly 2011, 2014).

[64] The promotion of a legal protection of the liberty to experiment in one's own body (biohacking, do-it-yourself biology, etc.) seems to find solid support among transhumanists. This preoccupation is not unrelated with the infamous corporative interest in owning genetic patents. Libertarian tendencies are thus always present among transhumanists. Indeed, the proactionary stance aims at counterbalancing libertarian attitudes that could end up absorbed by the very corporations that transhumanists want to challenge. For the development of the notion of morphological freedom see Bostrom 2011 and Sandberg 2013. For a proposal that would address this transhumanist concern by means of securing a legal framework of bodily ownership down to the genetic level, see Fuller 2013a, pp. 111–128. For a critical foundation of morphological freedom primarily based upon Modern social philosophy, see Fuller 2016.

[65] The notion was put together by Max More, in order to counter the precautionary stance that allegedly is the default epistemic attitude in the ethics of scientific research (More 2013b).

ground for converging technologies to take place and flourish, as the NBIC report of the previous section suggested. Transcending the traditional left/right dichotomy, a proactionary standpoint advocates for a risk-friendly approach while exercising state regulation for the sake of society's advancement and safety—against a position where one ought to know first the risks involved before continuing research. A paradigmatic case scenario would be a Welfare nation promoting a risk-friendly attitude toward novel scientific research, but at the same time safe-guarding its subjects via an *ad-hoc* insurance system. One could see here a libertarian attitude conflated with an emphasis on social support. Still another instance would be approaching the military (due to its inherent long-term goal mentality) to get funding for disruptive research that would ameliorate not only individuals but a centralized society as a whole. In this case one could identify the famed conservative "military-friendly" mindset merged with a socialist framework. In any case, this proactionary imperative might be the latest historical installment that exemplifies how the left/right divide becomes increasingly untenable.[66]

Transhumanist impulses, as stated above, can be identified in various instances throughout the history of philosophy. It is not the purpose of this work to give an exhaustive historical investigation of all transhumanist expressions throughout the evolution of Western thought.[67] However, some historico-philosophical instances seem to strongly underpin late scientific methodologies crucial for converging technologies to advance. These novel scientific attitudes may, in the view of this writer, help the cybernetic worldview to flourish this time. Facing this possibility, it is fitting to articulate some of the transhumanist impulses that conformed the metaphysical framework behind these technologies, since they might dissolve the very conundrums that brought the cybernetic enterprise to its demise. More importantly perhaps (particularly in what regards our acceptance of these technologies), these philosophical signatures are far from

---

[66] Steve Fuller has substantially articulated and expanded the proactionary concept into a book (Fuller 2013a).

[67] A full historical treatise of the historical sources of transhumanism remains to be written. However, having in consideration the "uplifting" core of transhumanist impulses, a good point of departure could be David Noble's *The Religion of Technology*—where the historic human tendency for elevating itself toward divine perfection, my means of technology, is treated (Noble 1997).

strange or alien to the Western tradition of the humanities. They indeed conform the very core of whatever it is that we call "civilization". These metaphysical skeletons, transhumanist in nature, may help converging technologies to realize the cybernetic project. Thus, these historical instances (one from the Medieval times and the other two from Modernity) will roughly correspond to the aspects shown above to be essential for cybernetics' ultimate fate; namely, the relevance of materiality for the nature of a machine, the mechanization of machine-unfriendly realms and constructability as a measure of epistemic success.[68]
Let us take a look at them.

SCOTIAN CONTINUUM BETWEEN THE MATERIAL AND THE IMMATERIAL

After several centuries of having Aristotle's works unavailable to Europe, the thirteenth century witnessed a revival of Aristotelian philosophy. After a millennium of a Christian philosophy largely based upon Platonism—inheriting at times its subsequent relative contempt for the physical realm—Saint Thomas Aquinas (1225–1274), a member of the newly formed mendicant Order of Saint Dominic, embarked upon the major task of relocating the fundamentals of Christianity upon the newly rediscovered Aristotelian writings. This displacement created waves of tension throughout Christendom. There were deep concerns regarding the outcome of combining the Christian faith with a philosophy which, even if acknowledging the existence of the Divine, it did it in a way that precluded this being from a relational intimacy with humans. Aristotle's god was not a personal god; even less one that would become man and dwell among us.

A prominent figure who relentlessly attacked Aquinas was Blessed Duns Scotus (1266–1308), a member of the order founded by Saint Francis of Assisi. Today both Doctors of the Church (*Doctor Angelicus*—Aquinas—and *Doctor Subtilis*—Scotus), their orders were founded just a century apart. They embodied a profound reform of the Church at the time, accentuating poverty as a charisma and quickly

---

[68] Chapter 7, Sections "John von Neumann's Appropriation of McCulloch and Pitts' Networks","The Letter to Wiener: Post Hoc, Ergo Propter Hoc?" & "The Hixon Symposium: Strengths and Weaknesses of the McCulloch & Pitts' Networks" respectively.

growing large all over Europe. During the lifetime of both thinkers the mendicant feature of their orders gave increasing space to an emphasis on education, eventually founding universities with their own distinctive character. Although the famed rivalry between the Franciscans and the Dominicans might have been somewhat exaggerated (the former being somewhat prone to adventurous thinking and the latter to orthodoxy), it is widely recognized that each order characterized the two major flavors of European universities for centuries to come. One under the light of an Aristotelian realism (Dominicans), the other under a more Platonic take on reality—with heavy leniencies toward mysticism (Franciscans).[69]

Scotus was concerned about Aquinas' understanding of concepts predicated of God's divine attributes. For Aquinas, when we speak of God as, for example, being good, we actually mean this "goodness" in an equivocal manner, not in the same way in which we use it when we refer to, for instance, people. The goodness of God can be only analogically related to that of humans, of creation, of health and so forth. Aquinas clearly emphasized the transcendent nature of God: even if there are cues in nature that can eventually lead us to Him (as in Aquinas' own *Quinque Viae*), we ultimately know more about what He is *not* than what He is.

Scotus had a problem with that. For him, negative knowledge was ultimately not knowledge at all. He feared that if Aquinas had it his way, the human capability for reaching God would be truncated. Furthermore, Aquinas would think of "being" as not necessarily entailing "existence". God *is*, which is more than to merely *exist*, since "existence", being contingent, necessitates a "being" to render possible this "existence" in the first place. For Scotus, this equivocality could have an epistemically catastrophic effect, given the nature of human intelligence. The primal object of the human intellect is "being inasmuch as being" (*ens inquantum ens*). That entails that if something is, it will necessarily fall within the realm of the intelligible. There is no being without existence. To affirm the contrary would allow for the possibility of a being beyond intelligibility, which could lead to the denial of existence of the Ultimate Being. Hence, for Scotus there is a gapless flow in our predication of the attributes of a being, which extends all the way to the inclusion of an immaterial, divine being. Otherwise, there could be the case of a being, including this

[69] Fuller 2011, ch. 2.

immaterial being, equivocally referred to, and thus removed beyond the grasp of our intellectual might. In his own words:

> And lest there be a dispute about the name "univocation", I designate that concept univocal which possesses sufficient unity in itself, so that to affirm or deny it of one and the same thing would be a contradiction. It also has sufficient unity to serve as the middle term of a syllogism, so that wherever two extremes are united by a middle term that is one in this way, we may conclude to the union of two extremes among themselves.[70]

For Scotus, in predicating, for instance, the "cleanliness" of a man, we can refer to the moral aspect of his persona or to his bodily realm—among other levels of qualified existence. So, one can say that such man is clean and not clean, without committing contradiction, given that one is applying cleanliness to different aspects, and still maintaining the univocity of being (*univocatio entis*) regarding cleanliness. The meaning of the noun (or adjective) is the same for both cases, but the variation occurs in the way or degree in which this notion is applied. We say about God as being good in an infinite way, but of creatures as being good in a finite way. That is why for the *Doctor Subtilis* when predicating of God's attributes, "wherever two extremes are united by a middle term that is one in this way, we may conclude to the union of two extremes among themselves".

The same goes for *being* in general: the radical opposite to nothingness is being, even if this being can be infinite (God) or finite (anything else). Regardless of the way in which being opposes to nothingness, the same notion of being is at work here, either applied to God or His creatures. This ontological continuum[71] should warrant our finite capacity of reaching the Infinite One. After all, our knowledge would differ from the Creator's in a manner of degree, not kind—pace Thomas Aquinas and his followers, some of whom later played a substantial role in the doctrinal tribunals of the Inquisition. Despite its alleged heretic potential, the

---

[70] Scotus 1300, p. 20 (*Ordinatio* I, distinctio. 3, part 1, quaestio. 2, numero 26).

[71] Certainly the idea of a *continuum* as the default state of affairs of reality was already perused by the Ancient Greeks. In the realm of the inanimate, it was understood that *natura abhorret vacuum*. Among living entities, a careful continuity in "how rightly Nature orders generation in regular gradation" was recognized by Aristotle (*De Generatione*, pp. II, 1).

Scotian bridge between the physical and the non-physical was thus established, and it fact it was subsequently accepted by several philosophers and theologians within the Church (e.g., by the Jesuits).

BACONIAN SUBSUMPTION OF NATURE
TO THE "MECHANICAL ARTS"

Two centuries later we can begin to see the theoretical roots for the Scientific Revolution in the work of Francis Bacon (1561–1626). Immanuel Kant tellingly opened his *Critique of Pure Reason* with a quote from Baron Verulam, Bacon's honorific title. Kant does this in order to highlight Bacon's role in the *new science*, having "made the proposal that partly prompted this road's discovery, and partly—in so far as some were already on the trail of this discovery—invigorated it further".[72] Strong words that find echo in the copious amount of scholarship on Bacon, the alleged father of the scientific method in early Modern science.

Ironically, an aspect of Bacon's work that has received relatively little attention was perhaps the main reason behind his construction of the "new science" in the first place—namely, the retrieval of the lost epistemic abilities that humans were endowed with in virtue of having being created in *imago Dei*.[73] Such relative absence in the scholarship is all the more surprising once we realize that "for Francis Bacon, the reform of learning was not a secular pursuit, but a divine mandate".[74] Indeed, he did not mince words for clearly pointing out that Original Sin catastrophically

---

[72] Kant 1787, B xiii.

[73] Peter Harrison extends the consequences of the awareness of a doomed human cognition due to Original Sin beyond Bacon's motivations for his new science. For Harrison, it actually shaped the whole of scientific inquiry in Modernity:

> [T]he biblical narrative of the Fall played a far more direct role in the development of early Modern knowledge—both in England and on the Continent—than has often been assumed . . . competing strategies for the advancement of knowledge in the seventeenth century were closely related to different assessments of the Fall and of its impact upon the human mind (Harrison 2002, p. 240).

[74] Mathews 2008, p. 51.

affected man's intellectual capabilities, removing the mastery over nature that Adam once enjoyed.

In *The New Organon*, Bacon asserted that "[f]or man by the fall fell at the same time from his state of innocence and from his dominion over creation".[75] The allusion to Genesis 3 is clear in his understanding of the dramatic loss we underwent due to Adam's fault: The Devil, represented in the book of Genesis by a serpent, introduced darkness and confusion in the human mind forever. In *The Great Instauration*, the sharp decrease in our intellectual capacities is sorely lamented by Bacon in a Trinitarian prayer, which asks for "illumination", one of the Gifts pertaining to the Holy Spirit: "Lastly, that knowledge being now discharged of that venom which the serpent infused into it, and which makes the mind of man to swell, we may not be wise above measure and sobriety, but cultivate truth in charity".[76] Indeed a grim depiction of our epistemic landscape is laid out by him.

However, there is a way out from cognitive darkness. Again in the *New Organon*, after referring to both our losses—of innocence and of mastery over nature—triggered by Original Sin, Bacon gives a glimpse of hope: "Both of these losses, however, can even in this life be in some part repaired; the former by religion and faith, *the latter by arts and sciences*".[77] Our intellectual capacity is corrupted, that much is clear. Hence, we need a way to put it in check, to rely on something solid. Reliance on our "natural lights" is out of the question, due to our inherited state of cognitive contamination. The way to get rid of the "idols"[78] of the mind will thus entail a return to hard reality. This will not be an easy feat and will not come "naturally". The Ancient Greeks were faithful realists, but their appreciation of nature, even with powerful insights, did not produce growing knowledge. Their experience as great philosophers, but certainly *not* great scientists, stands as a warning beacon for us:

[75] Bacon 1620a, p. 189.

[76] Bacon, 1620b, p. 74.

[77] Bacon 1620a, p. 189. Italics added.

[78] Idols of the tribe (race), of the cave (individual), of the marketplace (language) and of the theatre (authority) (Bacon 1620a, pp. 95ff).

Signs also are to be drawn from the increase and progress of systems and sciences. For what is founded on nature grows and increases; while what is founded on opinion varies but increases not. If, therefore, those doctrines had not plainly been like a plant torn up from its roots, but had remained attached to the womb of nature and continued to draw nourishment from her, that could never have come to pass which we have seen now for twice a thousand years; namely, that the sciences stand where they did and remain almost in the same condition, receiving no noticeable increase, but on the contrary, thriving most under their first founder, and then declining.[79]

What is the missing element then? Bacon, who was also a lawyer, was allegedly inspired by the torturous questioning of witnesses in court—which often provide surprising and fruitful outcomes. Indeed, we have to observe nature, but not as a passive gatherer, accumulating data and speculating about possible connections and causes. Rather, we have to make up situations that would force the studied object to behave in such way that we could satisfy our previously construed questions. Immanuel Kant, who appropriately understood the Baconian dictum, captured well Bacon's suggestion for pressing nature. In Kant's words, "reason must indeed approach nature in order to be instructed by it; yet it must do so not in the capacity of a pupil who lets the teacher tell him whatever the teacher wants, but in the capacity of an appointed judge who compels the witness to answer the questions that he puts to them".[80] Making allusion to the Ancient Greeks one more time, this time specifically to Aristotle, Bacon pointed out the crucial difference it makes to "push" nature into fitting inquiring categories, declaring that "in the business of life, a man's disposition and the secret workings of his mind and affections are better discovered when he is in trouble than at other times, so likewise the secrets of nature reveal themselves more readily under the vexations of art than when they go their own way".[81] Thus, the Ancient Greek experience, in contrast...

is the opposite of what happens with the *mechanical arts*, which are based on nature and the light of experience: they (as long as they find favor with

[79] Bacon 1620a, p. 113.
[80] Kant 1787, B xiii.
[81] Bacon 1620a, p. 130.

people) continually thrive and grow, having a special kind of spirit in them, so that they are at first rough and ready, then manageable, from then onwards made smoothly convenient by use—and always growing.[82]

Our fallible intellectual capabilities get anchored in physical substrata, but not by imposing its faults nor by letting itself be paved upon by nature's might. Instead, a reciprocal relation is formed, whose results let themselves get noticed by a growth of knowledge. When this constructive conflict is not present in the inquiry, the fate of the Greeks stands as a warning. The "mechanical arts" provide a key element of solidity, whose determinism (in so far as mechanical) acts as a reliable platform for our otherwise weakened reason to perform its questioning tasks. The notion of *experimentum crucis*, which constituted a methodological pillar for subsequent developments in science, was used as a basic element upon which much of scientific thinking regarding reliable parameters for experimentation was done. The set of interrogations that would construct the backdrop for the devised experiment would all constitute, once the results are obtained, a model of a portion of reality—the phenomenon's identified patterns of reaction to our experimentation.

The nature of the Baconian model generated itself inquiries, and its constitution, aims and even struggles were pondered in the centuries to come. Although Bacon effectively initiated a mechanical reconstruction of nature, the aspect of buildability that a mechanical model inherently possess would go on to inform realms that were perhaps not foreseen by Bacon himself, as we will see below. But first, let us explore one more instance of the evolution of this "mechanical epistemology", this time coming from one of Bacon's most fervent admirers—again, two centuries later.

VICONIAN CONSTRUCTABILITY AS A CRITERION OF TRUTH

Giambattista Vico (1668–1744) is perhaps best known for *La Scienza Nuova*,[83] a treatise on how to lead nations, later regarded as a foundational text for philosophy of history. His early writings, however, provide valuable evidence for establishing how dramatically important were Bacon's insights for the era. In an autobiography where the Italian philosopher candidly refers to himself in the third person, he speaks of his discovery of

---

[82] Ibid., p. 113.
[83] Vico 1744

the English thinker. He recounts that "from his *De augmentis scientiarum* Vico concluded that, as Plato is the prince of Greek wisdom, and the Greeks have no Tacitus, so Romans and Greeks alike have no Bacon".[84] Vico, himself a Catholic, castigated Baco's vicious attacks against the Church of Rome. However, trying to balance things out, he gave the hostile Anglican philosopher its fair due, saying of him that "without professional and sectarian bias, save for a few things which offend the Catholic religion, he did justice to all the sciences".[85] Vico indeed recognized Bacon's original contribution to scientific methodology, which would itself later reverberate into an augmentation of empirical knowledge.

Vico's metaphysics, developed years before *La Scienza*—and upon which the latter was based—had a strong philological accent. He advanced the argument that in the case of the Latin language, key linguistic terms had been formed out of philosophical insights rather than vulgar trades— indeed "derived from some inward learning rather than from the vernacular usage of the people".[86] Vico engaged into etymological archae- ologies to bring into the open the deep meaning of certain metaphysically dense concepts. As a way of guiding examples, he brought up the verb *intelligere* (which means both "to read perfectly" and "to have plain knowledge"); the verb *cogitare* (meaning both "to think" and "to gather"); and the noun *ratio* (which means the recognition of mathema- tical proportions as well as a man's endowed reason). Most tellingly, *verum* and *factum* were at the time convertible (interchangeable), as it is shown in the Latin adagio *verum esse ipsum factum*: "the truth is what is made, itself"—or, the truth is precisely what is done.[87] This assertion carries some immediate epistemological and ontological consequences. For starters, the only one who can *fully know* is God, for He—and He alone—creates its own object of knowledge.

---

[84] Vico 1731, p. 139.

[85] Vico 1731, p. 139.

[86] Vico 1710, p. 37.

[87] In Vico's words, "For the Latins, *verum* (the true) and *factum* (what is made) are interchangeable, or to use the customary language of the Schools, they are convertible" (Vico 1710, p. 45).

Before casting Vico's etymology-based metaphysics under a negative light, one ought to be reminded that he is not alone in this understanding of divine knowledge as the sole one which both knows something perfectly *and* creates precisely that something while knowing it. Indeed, Vico's views are not dissimilar from those of Immanuel Kant in this respect. Roughly a contemporary of Vico, Kant had a clearly exclusive adscription of the faculty of "intellectual intuition" to God alone. For him, although we are endowed with "sensible intuition" (space and time being its pure forms), our faculty is passive, receiving raw sensible material from the thing as it is presented to us. All of which would construct the uniquely Kantian understanding of "experience", ordered by our intellect. This is part of the core of the "Trascendental Aesthetic" in the *Critique of Pure Reason*. Our *intuitus derivativus* does not create the object of our intellect. In contrast, in the case of God's *intuitus originarious*, to think about something *is* to create that something:

> Our kind of intuition is called sensible because it is *not original*. *I.e.*, it is not such that through this intuition itself the existence of its object is given (the latter being a kind of intuition that, as far as we can see, can belong only to the original Being)...intellectual intuition seems to belong solely to the original Being, and never to a being that is dependent as regards both its existence and intuition (an intuition that determines that being's existence by reference to given objects).[88]

Vico's take on truth as convertible to that which is built, is mindful of Bacon's emphasis on our need for reliance on experimentation for extracting the answers from nature. However, the line of thought that goes from Vico's claim regarding the *verum factum* to Bacon's need for experimenting with reality, needs some explanation. God knows everything totally and perfectly because all things that constitute this "everything" are already in Him; humans, on the other hand, are external to these things—we have a superficial grasp of them, since they are not internal to us:

> God reads all the elements of things whether inner or outer, because He contains and disposes them in order, whereas the human mind, because it is

---

[88] Kant 1787, B 72.

limited and external to everything else that is not itself, is confined to the outside edges of things only and, hence, can never gather them all together.[89]

The nature of ultimate knowledge has a bearing on science "in general", namely, both for God and for humans. Science is "knowledge of the genus or mode by which a thing is made; and by this very knowledge the mind makes the thing, because in knowing it puts together the elements of that thing".[90] Scientific knowledge, thus, implies a creative process. One knows in the very act of creating. Of course, only God fully knows, since He creates. However, there is a realm where we also know while creating the very object that we know. That realm is mathematics: "Just as he who occupies himself with geometry is, in his world of figures, a god (so to speak), so God Almighty is, in his world of spirits and bodies, a geometer (so to speak)".[91] It is in mathematical knowledge where we get the closest resemblance to divine cognition.

The situation is dramatically different when it comes to the understanding of the natural order, that is, knowledge of physical things. Here, once again, both God and humans possess science, but ours is essentially different from His. We drag an inherent stain in our scientific endeavors regarding the natural realm—namely, our lack of ability to "create" (available to God alone) and our consequent need for abstraction. After all, "[s]ince human knowledge is purely abstractive, the more our sciences are immersed in bodily matter, the less certain they are".[92] However, Vico confided with us an aspect we ought to emphasize while doing science in order to overcome this genetic limitation: "Just as divine truth is what God sets in order and creates in the act of knowing it, so human truth is what man *puts together* and makes in the act of knowing

---

[89] Vico 1710, p. 46.

[90] Ibid.

[91] Miner 1998, p. 66.

[92] Vico 1710, p. 52. Recall Ashby's need to liberate the machine from physical substrata, indeed using the nature of geometry as his guiding example (Chapter 8,

it".[93] Constructability—"putting together"—is the first criterion of truth, and thus, of scientific success.

In this context, Vico reminded us that "God knows all things because in Himself He contains the elements with which He puts all things together. Man, on the other hand, strives to know these things by a process of division".[94] This "process of division" literally informs the essence of science, while occupying itself with the "anatomy of nature's works".[95] This "anatomy" goes beyond a mere figure of speech. We need to dissect, *qua* physician, the elements of nature, and then *cognitively*—not *existentially*, which is only divinely feasible—recreate the object according to our inquiry. Thus, when Vico refers to "dissection", he is indeed referring to experimentation.[96]

Vico's indebtedness to Bacon is more explicit when he makes allusion to an incipient Modern scientific method. He asserted that "hypothesis about the natural order are considered most illuminating and are accepted with the fullest consent of everyone, if we can base experiments on them, in which we make something similar to nature".[97] Considering the ingeniousness of Bacon's proposal pointed above—namely, by pressing nature to answer *our* questions—we actually are anchoring ourselves in reality, not removing our minds from it. This critical reverence toward reality begets *wisdom*, which ultimately "in its broad sense is nothing but the science of making such use of things as their nature dictates".[98] According to Vico, Bacon did nothing less than bridging our intellectual capacity with the natural world. However, it is Vico who carries the Baconian lesson to its fulfillment. The experimental reconstruction of a phenomenon under controlled conditions is just a first

Section "The Relevance of Materiality for the Nature of a Machine: From Complexity to Disembodiment").

[93] Vico 1710, p. 46. Italics added.

[94] Ibid., p. 48.

[95] Ibid.

[96] Miner 2004, p. 103.

[97] Vico 1710, p. 52.

[98] Vico 1744, CXIV: 326.

step toward the recreation of the object. But true knowledge is only acquired when the model becomes the modeled![99]

This closes the Scotian circle, where a type of strong relation between physical entities and disembodied realms is established. Vico's claim regarding "hard sciences" is telling in this respect: "The most certain sciences are those that wash away the blemish of their origin and that become similar to divine science through their creative activity in as much as in these sciences that which is true and that which is made are con-vertible".[100] Vico's prescient *verum factum* seems to neatly link Scotus' *univocatio* and Bacon's *mechanical arts* into a zenith of knowledge attain-ment that drives an important aspect of contemporary scientific research. It should, thus, not come as a surprise that one can find in our times a research laboratory in the Netherlands named "Giambattista Vico Institute for Cybernetics and Applied Epistemology".[101]

In fact, the whole rise and demise of cybernetics could be understood as having suffered such fate in an important degree due to its legendary emphasis on buildability as the core of its scientific methodology—itself deeply rooted in the drive behind post seventeenth-century scientific practice. The cybernetic tension occurring between, on the one hand a theoretical structure, and on the other, the physical anchoring of such theory (e.g. von Neumann's kinematic model), gets all the more interest-ing once one notices how contemporary science *still* understands its explanatory task in fairly Modern terms—namely, roughly mechanistically. These values are indeed essential to the core of the Western tradition, as expressed in the longing for a "divine" knowledge that entails the possi-bility of radically transforming (tinkering, fixing, and improving) a crea-tion given to us for our disposal—ourselves included. Such impetus for control and reconstruction is ever more present with us, and in tangible ways, as we shall see in the next chapter.

---

[99] The cybernetic maxim "the best material model for a cat is another, or preferably the same cat" (Rosenblueth and Wiener 1945, p. 320) can be seen now under a different light: Nested in the classical Western tradition of the humanities and philosophy.

[100] Vico 1710, p. 52.

[101] Jung 2005.

CHAPTER 10

Transhumanist Technologies: New Possibilities for a Cybernetic Worldview

CONSTRUCTABILITY AS A MEASURE OF EPISTEMIC SUCCESS: MOLECULAR REORDERING OF REALITY

Let us go back to the last moments of cybernetics. Only 5 years after von Neumann's latest attacks against McCulloch and Pitt's networks at both the Hixon Symposium and the Illinois lectures, a Manhattan Project colleague of his, Richard Feynman (1918–1988), better known later for his Nobel price research on quantum mechanics, would be refusing a teaching invitation from von Neumann's hiring institution—Princeton— despite having also earned his doctorate from there in 1942. Instead, Feynman accepted a teaching position at Caltech, where he delivered a talk that would come down in history as the first that addressed the possibility of manipulating reality at a level where the transformation and recreation of objects becomes scientifically feasible. The one-time lecture entitled "There's Plenty of Room at the Bottom"[1] remained, however, largely unknown. Almost three decades later, Eric Drexler, a doctor in engineering from MIT, found it and published it as part of his 1986 book *Engines of Creation: The Coming Era of Nanotechnology.*[2]

---

[1] Feynman 1960.
[2] Drexler 1986.

© The Author(s) 2017
A. Malapi-Nelson, *The Nature of the Machine and the Collapse of Cybernetics*, Palgrave Studies in the Future of Humanity and its Successors, DOI 10.1007/978-3-319-54517-2_10

Although the term "nanotechnology" was coined a decade earlier by Norio Taniguchi (1912–1999), it is after Drexler's book that the notion started to gain traction in the engineering, and later in the science, communities. The journal *Nature* has now a permanent section dedicated to research at the nanoscopic scale.

Canonically, nanotechnology is said to aim at the mechanical manipulation of matter at the molecular level—literally, atom by atom. Such level of reality is allegedly of utmost importance, since stuff can be still "classically" understood under a Newtonian light and rearranged without incurring in issues of quantum indeterminacy.[3] Since a physical thing owes its nature to its particular atomic arrangement (one modifies this arrangement and one ends up literally with something else), the possibilities immediately foreseen are mind-numbing. In principle, an eventual "molecular assembler" could transform any physical entity into another one, by means of reordering its atomic structure. Such an envisioned feat would obviously transform the world as we know it in radical ways, replacing its economy, resetting global health, and trivializing some deep questions—such as the possibility of artificial intelligence and artificial life (now amenable of being duplicated and enhanced, instead of "found" or "created").

Even if by consensus the field still is in its infancy, the first decade of our century has witnessed some relatively important achievements, which keeps fostering interest in the discipline in a gradual but relentless way. Feynman's explicit proposal, even if only theoretical, tried to garner attention toward that "untreated" level of reality in the midst of the Cold War. He predicted that research performed at the nanoscale[4] would have as beneficiaries, among other fields, microscopy and computation. Indeed, only 6 years later, in 1965, George Moore

---

[3] Quantum effects, however, are part and parcel of the whole nanoscale somewhat "exotic" halo. Within the threshold of 1–100 nanometers, the effects pertaining to the so-called "quantum realm" are still present, and thus, materials tend to behave in a way that is absent at the macro-level of reality (e.g., an otherwise inert material becomes conductive, etc.). For an addressing of the quantum-related problems in nanoscale research, see Drexler's "An Open Letter to Richard Smalley" in Drexler 1986, Appendix.

[4] The prefix "nano-" for these matters was not yet coined at the time (1960s), so Feynman did not use it.

would be stating what would be henceforth known as "Moore's Law": Every 18–24 months the number of transistors that could fit in an integrated circuit would double. Thereafter, somewhat expectably, two cornerstones in nanotechnology occurred in a common area shared between computation and microscopy. In 1981 two IBM scientists constructed a "scanning tunneling microscope", which for the first time allowed humans to see surfaces at the atomic level, awarding them a Nobel prize five years later.[5] Eight years later, another two IBM scientists rearranged at will 35 individual atoms, forming the logo "IBM". This last achievement properly showed that what nanotechnology was aiming for—the mechanical manipulation of atoms—seemed to be in fact feasible.[6]

As indicated previously, by the turn of the millennium the US National Nanotechnology Initiative was established, receiving its first funding during the administration of President Bill Clinton.[7] Quickly realizing nanoscience's potential for multilayered pervasive and disruptive consequences, in 2004 the European Union produced a document entitled *Towards a European Strategy for Nanotechnology*.[8] The European community made clear its preferred emphasis on the possible benefits for society as a whole, rather than on the human individual, as the American initiative would seemingly suggest. That same year, the United Kingdom's Royal Society released the report *Nanoscience and Nanotechnologies: Opportunities and Uncertainties*.[9] In this report Great Britain called for focusing on investigating the possible toxic side effects of doing research at the nanoscale.

By 2005, further manipulation of individual atoms was accomplished, this time by means of constructing a nanoscale "car". A team lead by James Tour (Rice University) put together four fullerenes[10] united by a "frame" composed of hydrogen molecules. The experiment was

[5] The awarded scientists were the German Gerd Binnig and the Swiss Heinrich Röhrer (Nobelprize.org 2014a).

[6] Eigler and Schweizer 1990.

[7] Chapter 9, Section "Cybernetics 2.0: The Nano-Bio-Info-Cogno Convergence"

[8] European Commission 2004b.

[9] Royal Academy of Engineering 2004.

[10] A fullerene—or "buckyball"—is a spherical carbon molecule.

designed to find out in which way fullerenes move across materials. When the surface (the "road") got warmed up (e.g., to 200 degrees Celsius), the diminutive cars began to run over the road at relatively "high speed". Although the "nanocar" did not actually have a motor,[11] the idea of a nanomachine, indeed proposed by Feynman as a future possibility, seems to be closer to reality. The 2016 Nobel prize in chemistry was conjointly awarded to three scientists "for the design and synthesis of molecular machines".[12]

Let us stop for a minute and consider the import of these advances. One could argue that these pertain to the very heart of scientific practice *tout court*. Vico's aim for an absolute recreation of the studied object (the model) as the only—or at least, the best—signature of true knowledge has certainly been at the heart of the scientific mind since the beginning of Modern science. Indeed, the case could be made that such aim *is* the essence of what is expected out of Bacon's *experimentum crucis*. Unfortunately, due to the nature of physical reality—leave alone the nature of human cognition—all we could aimed for was the best possible description of a *behavior* manifested by a certain phenomenon or object. However, as von Neumann realized, the mapping of a complex system's behavior triggers an explosion of variables, which spells doom for the set-up model. There comes a point where the range of possible behaviors is just too immense to be captured and articulated. It would seem that at that moment, doing a piece by piece mapping of the *structure* of the complex object itself becomes more feasible. Again, however, in light of the nature of both physical reality and human cognition, this has shown to be a practically unattainable feat, due to the staggering complexity of the object's structure—which made scientists focus on behavior in the first place (e.g., behaviorism's reduction of the human person to a black box). Ultimately, the endeavor for mapping and rebuilding the actual structure of an entity—a.k.a., attaining *truth*, in Viconian terms—was right from the start inherently beyond our reach. That is changing.

Von Neumann's plea for access to the structural detail of the studied object, "flabbergasting" at his time, might begin to seem more

---

[11] The nano vehicle lacked an engine, and thus, it was not a car in the full sense of the word (a.k.a., a machine with four wheels, self-propelled by a motor) (Shirai et al. 2005).

[12] The scientists awarded are the French Jean-Pierre Sauvage, the Dutch Fraser Stoddart and the Scottish Ben Feringa (Nobelprize.org 2014b).

accomplishable now. When he rejects the behavior-based model, in order to focus on the model of the object itself, he is still remaining within the realm of modeling—but this time the model would have a one to one, "piece by piece" correspondence to the actual object. In contrast, when we become capable of not only seeing parts that previously were only theorized about, but also of moving them around at will, the wish for entering the structure of the object seems to get closer to fulfilment. Von Neumann's locomotive imagery for dissecting very complex (and very small) objects would now seem to be nearer full instantiation. And just as no modern scientific task could be said to have a Scientific Revolution heritage without exhibiting a mechanistic explanation (euphemistically referred to as a "explanatory mechanism"), by the same token no contemporary scientific pursuit could be said to have cybernetic heritage without the presence of machines. And if that is the case, apparently, nanoscience is on its way to have them too.

With nanoscience, the epistemic hope of knowing the object in an almost "divine" way seems to be amenable of physical realization: The Viconian horizon of literally recreating our object of study as proof of true knowledge could finally be within reach. Taking us closer to attaining ultimate knowledge and control, the constructed model could become so close to the modeled object that any distinction would be rendered trivial. Vico's *verum factum* is apparently close to actual instantiation at a qualitatively distinct level. After all, the famous—if unfulfilled—cybernetic dictum was that "the best material model for a cat is another, or preferably the same cat".[13] Reordering atoms one by one, indeed restructuring the very fabric of reality, would have sounded to Vico as the logical step, after Bacon, toward the final understanding and mastery of nature.

MECHANIZING MACHINE-UNFRIENDLY REALMS: BLURRING THE NATURAL WITH THE ARTIFICIAL

The possibility of building nanomachines, beyond instantiating Vico's dictum, substantially advances the profoundly ambitious Baconian project—project that was recognized and praised by Vico himself. Not only can objects be partitioned atom by atom, but these parts can be so

---

[13] Rosenblueth and Wiener 1945, p. 320.

rearranged that a working *machine* could eventually arise. Let us recall the cybernetic criterion of feasibility: Nature was able to do it, so it is doable.[14] After all, that is what organisms are conformed of: countless nanomachines working in perfect synchronicity, making up what is regarded as a living system. In fact, the addition of such realities constitutes the totality of biology.

At this point, the distinction between the natural and the artificial has already collapsed—further confirming one of Ashby's main arguments.[15] The reckoning of the building blocks of reality subsequently allows for their manipulation and arrangement within a grand mechanical process. In nanoscience, both the organic and inorganic are in principle equally dissected, treated and put together at will. As the NNI mentioned above points out, scientists have been noticing that "a majority of biological processes occur at the nanoscale".[16] The width of the strand of the DNA helix is around 2 nanometers (nm); the average size of a virus is 40 nm; a small bacteria's size is 200 nm; and so on. One classically mechanistic case in biology occurring at the sub-cellular level is that of the ribosome, which strikingly resembles a factory assembly line—indeed a natural "molecular assembler".[17]

Biologists have in later years expressed interest in another instance in nature that offers an even more striking example of machinery functioning at a nano-level. The bacterial flagellum (the attached "tail"—cilium—that gives bacteria aquatic locomotion), has been of interest to biologists, philosophers and even theologians[18] for several reciprocally intertwined reasons, all amenable of being put under one conceptual umbrella. Due to progress in crystallography and nano-photography, we are now able to see as part of the organism what could only be referred to as an "off-board" motor. The flagellum itself, a tube usually 20 nm wide, is attached to a

---

[14] Since the machinery of nature found a way to do it, it is evidently feasible: Recall the exchange between Ashby and Pitts on this type of reasoning (Chapter 6, Section "The Macy Presentation: Keeping Cybernetics Honest").

[15] Chapter 6, Section "William Ross Ashby's Nature-machine Equalization".

[16] National Nanotechnology Initiative (nano.org), Nanotechnology 101.

[17] Drexler 1986, ch. 1.

[18] This organism is of utmost interest for those who look at instances of nature that allegedly cannot be explained via Darwinian evolution. More on this in the next section.

complex array of amino acids that make up a structure which, beyond resembling an engine, seems to *be* an actual engine. One can readily recognize pistons, a camshaft, levers, and an axle. The rotation produced (which allows the flagellum to act as a propeller) is analogous to the kinetics found in an engine: pistons moving an axle via the camshaft, which in turn makes an exterior propeller turn. Endowed with this acquired motility, the bacterium can freely navigate through a liquid environment. The organic system has found a way to achieve locomotion in a thoroughly mechanical manner. Once we have identified a naturally assembled motor, we have found the "proof" that nano-motors are in fact possible. Cyberneticians, ostensibly, would not be surprised.

There are several issues to consider here, starting with the adjudication of the notion of "machine" to the mechanism connected to the flagellum. At earlier times, when pictures of the structure were substantially blurred, there was debate regarding the possibility of an anthropomorphization taking place in whatever we were seeing. After William Paley,[19] we are uneasily aware of the problematic consequences of finding and proclaiming real machinery in nature. However, when the images became strikingly clear, via mastering the complete atomic structure of the area responsible for the bacterial motility, little doubt remained regarding whether we are witnessing an actual natural motor. We indeed are.[20]

One feature particularly relevant for this section stands out. Nobody seems to have a problem *now* with one premise which, in the past, would have likely stopped the dialogue in its tracks. Everyone seems to take for granted that what we have before our eyes is, beyond any reasonable doubt, a machine. Nature seems to have been lately further stripped from any surrounding aura of mystery, being thoroughly mechanized all the way down. From a cybernetic lecture of reality, there is considerable import in this assertion—or realization. Facing the bacterial flagellum, whatever vitalistic or exceptionalist remnants of a metaphysic tradition that were

---

[19] Paley 1802.

[20] The Dean of Frontier Biosciences at Osaka University, Keiichi Namba—a researcher thoroughly committed to bacterial flagellum studies—does not hesitate in identifying in such a mechanism an actual engine: "The bacterial flagellum is a rotary nanomachine that spins at hundreds of revolutions per second driven by the electrochemical potential difference across the cytoplasmic membrane" (Namba 2010, p. 417).

seemingly still at work in the most materialist thinkers at the time of cybernetics (mid-twentieth century), seem to be extinct toward the second decade of the tweenty-first century. The metaphysics-mongers are no longer here. Probably just a few of us today would go through any pain in acknowledging that what we are witnessing is *both* a product of nature *and* a machine. A fully natural and fully mechanical entity. This is probably the most typically cybernetic feature of all the ones we have—somewhat unwittingly—inherited from cybernetics proper.[21] And it points to something deeper than just a biomimetic venue for future nanoscale research.[22]

As Heidegger was quick in pointing out, once we understand a phenomenon, it is framed in such way that we will, once we have the means, transform it and reorder it at will.[23] The mechanization of reality is well beyond recognizing patterns of machinery in nature, both behaviorally and structurally. In 2007, Craig Venter isolated the DNA code of a species of bacteria (*Mycoplasma mycoides*) and then transferred it to another species (*Mycoplasma capricolum*). This resulted in the latter species behaving like the former one, effectively amounting to a "change of species"[24]: The equivalent to a car "chassis" to be later used in different recombinations at will.[25] The next year, Venter sequenced the complete genome of the bacterium *Mycoplasma genitalium*, then stripped it of ¼ of its DNA (i.e., removing its pathogenic features), reassembled it with identifying "watermarks", and produced several clones—which exhibited the introduced characteristics (including the watermarks). The "new species" was named *Mycoplasma laboratorium*.[26]

---

[21] Fully fleshing out the lineage between cybernetics and current scientific disciplines would entail a pain-staking articulation that remains to be done. As indicated in Chapter 3, Section "The Decline", Margaret Boden's two-volume work attempts to fill this gap in what pertains to, at least, the sciences of cognition (Boden 2006).

[22] Keiichi Namba believes that the "structural designs and functional mechanisms to be revealed in the complex machinery of the bacterial flagellum could provide many novel technologies that would become a basis for future nanotechnology..." (Namba 2002).

[23] Heidegger 1977a.

[24] Lartigue at al. 2007.

[25] ETC Group 2007, p.3.

[26] Gibson et al. 2008.

Venter worked once again in 2010 with *Mycoplasma mycoides*, this time synthesizing its DNA, storing it in a computer, and then inserting the digital footprint in *Mycoplasma capricolum*. The resulting "bacterium", nicknamed "Synthia" (after the name of the synthetized version of the donor bacterium, *Mycoplasma mycoides* JCVI-**syn**1.0), was able to "normally" reproduce, showing features of the *Mycoplasma mycoides*—not of *Mycoplasma capricolum*.[27] It effectively swapped its identity. By 2016, Synthia has been stripped down of "useless" DNA so that it can now reproduce in a matter of hours, not weeks—becoming *Synthia 3.0*.[28] Although the new synthetic organism is still relying on elements that are not abiotic (i.e., it still uses the bacterium's own reproductive machinery), the aim is to later construct an entire bacterium, and eventually a cell, synthetic in its entirety. In *late* cybernetic parlance (the one that brought tension to the project),[29] we do not only want it to behave as we intend it to, but to structurally *be* what we wish it to be.

It would seem that not only von Neumann's truncated efforts are now becoming doable. Ashby's notion of a machine—where the material of which is constituted (or lack thereof) is deemed irrelevant for its being—is permeating science beyond the complete blurring of the natural with the artificial. Indeed, there is another realm which lay at the core of the cybernetic conundrum, which in virtue of its identifiable transhumanist potential, may now produce a kind of knowledge validation that cybernetics could not afford. As an epilogue to the technologies that will avert the issues behind the cybernetic implosion, let us see it in the next section.

THE RELEVANCE OF MATERIALITY FOR THE NATURE OF AN OBJECT: SIMULATION AS SCIENTIFIC *FACTUM*

The context where this technology is introduced may appear prima facie strange. Later on, this very context will point to the importance that this technology, which underpins a relatively new scientific methodology, is having in contemporary scientific practice at large.

---

[27] Gibson et al. 2010.

[28] Hutchison III et al. 2016.

[29] Chapter 8, Section "Constructability as a Measure of Epistemic Success: The Intractability of a Highly Complex Model".

During the fall of 2004, the public high school of the town of Dover, Pennsylvania, introduced a new biology textbook[30] where Neo-Darwinian evolution was presented alongside an "Intelligent Design" (ID) alternative account. Following a mandate from the school board, a statement had to be read out loud by the teacher during science class, referring clearly to the existence of both accounts—evolution and ID—so that the student could evaluate and decide on its own. By next fall, a group of concerned teachers and parents had initiated a law-suit against the school district. The plaintiff sustained that the school board mandate amounted to a surreptitious introduction of religion into science class at a public school—which, following the idiosyncratically American "separation of church and state", would be against the law.[31] It soon became clear that what was at stake was the country's legal standpoint regarding science's identity.

Expert witnesses were called to testify. Those friendly to a broadly understood version of ID emphasized the religious motivations behind the Scientific Revolution. As we saw earlier, this event can be understood, if we pay attention to Francis Bacon's open motivations, as a tension within Christianity in order to overcome our "genetic" epistemic impotence set off by Original Sin.[32] In this context, Isaac Newton could be said to have held some version of ID.[33] The methodology of science, indeed the "scientific method", was brought up. There were claims that ID had a valid space in scientific discourse based upon this very method. ID supporters denied the interference of any preconceived theological element, allegedly relying instead on methodological deduction, hypothesis construction and subsequent induction following the evidence.

ID theorists traditionally try to capitalize on Darwin's own clause regarding the weakness of his whole theory should cases unexplainable by natural selection would arise.[34] Following this strategy, instances such

---

[30] Davis and Kenyon 1989.

[31] For a detailed account of the whole trial, see Humes 2007.

[32] Chapter 9, Section "Baconian Subsumption of Nature to the "Mechanical Arts"".

[33] For an account of the place of ID in the history of science (in relation with the trial), see Fuller 2008.

[34] "If it could be demonstrated that any complex organ existed, which could not possibly have been formed by numerous, successive, slight modifications, my theory would absolutely break down. But I can find out no such case" (Darwin 1859, p. 189).

as the above mentioned bacterial flagellum were presented in detail, purportedly showing in strictly scientific terms that there is evidence for the existence of "irreducibly complex" machines in nature—dynamical arrays whose perfectly fitting internal parts cannot be accounted for by evolution. It was claimed that these "parts" need to have been "intelligently designed" to fit each other and function in tandem. And that *that* is not how evolution works. Therefore, Darwin's self-effacing clause, so they claimed, would apply.

As it could be expected, the criticism against the ID arguments was relentless. The evolutionary counterarguments ultimately pointed to the alleged ID's undisclosed but defining character—a pseudo-scientific religious creationism operating in disguise. However, just as ID was being characterized as a non-scientific duplicitous view, it became gradually clear that, in all fairness, "evolution" had to be itself submitted to the same judging scope. The taken-for-granted "scandalously obvious" truth of Darwinian evolution had to be somehow *shown*. How do we know, with the sharp certainty that only science can provide, that evolution by natural selection actually obtains? After all, ID was being attacked from those very grounds: its purported scientific character was being pulverized. Accordingly, US District Judge John E. Jones III—a Republican practicing Christian—asked for the same sort of evidence that evolution advocates were claiming ID cannot produce.

Robert Pennock, a philosopher of science and expert witness in the trial for the side of evolution, came up with what he claimed was proof that neo-Darwinian evolution is *not* "just a theory".[35] Pennock brought a computer machine and put it on display in the courtroom. Avida[36] is a computer program that generates a virtual "evolutionary" environment. More specifically, the program is designed to create and maintain entities that "compete" for resources. Each has its own memory allocation, virtual

[35] The perceived exchangeability between the notions of "hypothesis" and "theory" seems to occur mainly in the English language. The statement of evolution as being "just a hypothesis" would in any case make more sense. Evolution as "just a theory" can be translated to evolution as "just a *tested* hypothesis"—the canonical definition of theory—which would be of course problematic. For the problems this inaccuracy raises in science teaching, see Williams 2013.

[36] For an explanation co-written by Pennock of the role of Avida in scientific methodology, see Lenski et al. 2003.

central processing unit (CPU) and "protection" from each other. They evolve in such way that they can gain access to time with the CPU. It is worth noting that the "design" element is exhausted at the creation of the program itself; it does not extend to its procreated "entities". These in fact behave autonomously. The digital beings have the capacity to modify (re-program) themselves, in order to reach a better fit for survival—not unlike sophisticated computer viruses that fight for their "lives" in domestic computing environments. Pennock is aware of the epistemic consequences that this digital system entail, claiming that "it is clear that natural selection can be perfectly instantiated in A-life systems".[37] Furthermore, in his view, "this is not a simulation"[38] of evolution—it is an "instance" of it. Materiality is not deemed necessary for accepting the truism of evolution:

> While it is true that evolution of carbon-based organisms is the prototype of the concept, this historical fact is not a sufficient reason to limit its scope. Why be carbon-centric? It is the patterns of causal interactions that are relevant, not the particular *material substrate*....the material substrate of the Darwinian processes should be *irrelevant* to whether we recognize something as an instance of Darwin's evolutionary mechanism.[39]

The attentive reader will have noticed that the above words could have been naturally uttered by Ross Ashby 65 years ago. Now, in regards to an important ruling that is set to affect the way science is taught in North America in years to come, they seem like an uncanny confirmation of what he suggested: Scientifically—as in mechanistically—speaking, materiality is irrelevant for an object to be what it is. Back then it was for a machine. This time it is for evolution—phenomenon underpinned by mechanistic processes anyway. Both are real, even if materiality is not part of their nature *per se*. The actual case, the material one, is none but one instantiation, among many other possible ones. Both the immaterial and the physically material are equivalently real. And if we bring up Vico, let us recall, the immaterial version would be even "truer", since it may encompass more instances, while the material one embodies just *one*.

---

[37] Pennock 2007, p. 37.

[38] Ibid.

[39] Ibid., p. 32. Italics added.

For Pennock, Avida showed that Darwinian evolution is a falsifiable theory amenable of being discarded and superseded—in other words, a tested hypothesis. A scientific theory. Which happens to be, in this case, confirmed. Even if the witnessed "evolution" happened entirely inside a machine, one nevertheless can claim now to have "proof" that evolution is, plainly and simply, true. Judge Jones was convinced. The court declared evolution a *fact*. And the defendants lost.[40]

Facing an instance of the awkward relation between scientific practice and institutionality, one could object that a legal decision cannot make, let alone guide, science. This objection could see in this court ruling, at best, a change of attitude regarding what is real and what it is not in legal environments (e.g., more virtual facts are counting as legal realities, as in the case of online extramarital affairs). However, once we factor in Pennock's statements, we can readily notice that this epistemic ethos, beyond contemporary legalisms, is increasingly shared by salient scientific enterprises. Virtuality becoming reality, models becoming the object, are becoming ubiquitous. Indeed, Pennock is not alone in advancing these views—and for what is worth, the judge probably knew it. Some relatively recent examples are relevant.

After almost six decades that von Neumann complained to Wiener about the impossibly daunting task that he was reluctantly suggesting as cybernetics' next frontier (constructing models of actual structures, instead of those of behaviors), the task seems to be now closer to feasibility—with a twist. Advances in microscopy, crystallography, nanotechnology, and indeed the very field that he helped launch, computer science, are allowing for experimental tools that open up new vistas in scientific methodology and practices. In fact, pioneering science seems to be possible due to these novel technologies employed.

---

[40] To be sure, the ID side (the defendants) lost the case not only due to this "demonstration". An early printed draft of the controversial textbook was found by investigators, and in it, the word "creationism" was scratched and replaced with "intelligent design" on top. For the judge, this was incontrovertible evidence that religion was indeed being smuggled into science class in a public school, thereby violating the constitutional separation between church and state in the United States.

In 2006, a group of scientists lead by Klaus Schulten (from Ashby's own University of Illinois at Urbana-Campaign) simulated an entire virus, modeling it atom by atom. Von Neumann would have been delighted by the fact that the study was published in the journal *Structure*. The satellite tobacco mosaic virus, one of the smallest in existence, had its approximately one million atoms "recreated" simultaneously. It required the sheer computing capacity of one of the most powerful machines on the planet, housed in the National Center for Supercomputing Applications, located at the same university. Still, it took 100 days to bring up to "life" only 50 nanoseconds of the "virtual" virus.[41]

Those frowning upon the fact that recreating a virus does not necessarily mean recreating a living entity, could do so for only 6 years. In 2012, Markus Covert (from Stanford University) simulated an entire living organism—a bacterium.[42] This unicellular entity, the *Mycoplasma genitalia*, is the living being with the smallest genetic footprint: 525 genes. Remarkably, this time the entity recreated *in silico* underwent a reproductive process. Serving as a witness to the kind of progress that computation can develop in just 6 years, the virtual entity lasted 10 hours: The actual time the modeled entity takes to reproduce—and remarkably, the exact time the constructed model took to do it as well. The constructed model and the modeled organism were operationally identical, all the way down to their atomic existence. For attaining this, a cluster of 128 powerful computers interconnected were required. It was the first time a complete organism has been completely reproduced inside a virtual environment.[43]

In all fairness to cybernetics, these scientific accomplishments were the stuff that made up the cyberneticists' dreams—albeit unaccomplishable at their time, due to the lack of technological advancement in what regards computing power and manipulation of materials.

Von Neumann's one-to-one mapping of the model to the modeled is seemingly being accomplished, with one caveat: It is not happening in three-dimensional space, but in a disembodied realm. Perhaps thanks to Ashby, and if plunging deeper, to Scotus, it would seem that science these

---

[41] Freddolino et al. 2004.

[42] It is traditionally understood that a virus does not qualify as a "complete" organism, given that it needs a host to survive. In fact, debates regarding its status as a "living" entity pivot upon this very issue.

[43] Karr, Sanghvi et al. 2012.

days does not necessarily subscribe to the cybernetic dictum of buildability as epistemic success. The context of what is scientifically real has changed since the 1940s of classical cybernetics. Simulations now count as both the evidence and object of study.

However, if one were to recognize the positive aspects of the Viconian mandate of constructability present in the cybernetic ethos, we *also* have that. Nanotechnology aims at the atomic reordering of reality itself. Furthermore, advances in nanoscience may point to the possibility of a molecular assembler, in which case the cybernetic organization of reality could become indeed accomplishable: Nature, being in essence mechanical, can be tinkered, deconstructed and constructed again at will.

The cybernetic tension between the mandate for material model construction and the seemingly perennial temptation for abandoning all grips with physical reality since Plato, might have proven to be deeper and more pervading—and might had spelled, at its time, the demise of the cybernetic project. In face of the novel technologies depicted above, some relevant questions arise: To what extent this tension would be ameliorated, or even dissolved, in our days—at least in what refers to the conflict between modeling a structure versus a behavior? What is the import of this resolution for current instantiations of the cybernetic metaphysical programme? What is the impinging of all this upon the advance of science and the consequent modification of the human condition?

We came a long way since von Neumann realized the impossibility, to his own chagrin, of modeling an object's structure itself. The means to accomplish such a feat at the time was beyond anyone's reach. Now we can do something that falls squarely within von Neumann's methodological philosophy, thus somewhat dissolving the crippling issue of implementing the approach. It would be short-sighted to not recognize a cybernetic impetus in the contemporary accomplishments in this respect. Cybernetics did spouse a mandate for model-building in reality. But now the situation is that the definition of the "real" is undergoing transformation—fluid situation where we can also see a cybernetic signature, via Ross Ashby.

We can now satisfy both conflicting cybernetic requirements that back in its day signified an unsolvable internal tension. With super-computer simulation, a disembodied model can actually become the studied object and we are okay with that, given the extreme precision with which such model was put together. On the other hand, if we still demand a material instantiation of such model, we can also provide it, given our inroads into nanotechnology. All these underpinned by perhaps the instantiation of the most cybernetic of all influences: the attainment of mastery over creation by means of the actual equalization between nature and machine.

CHAPTER 11

Concluding Remarks

The main aim of the preceding chapters has been to establish the strong correlation between the evolving role of the notion of a machine, its role in scientific explanation and the subsequent outcome in the scientific pursuits it fostered, with particular emphasis on cybernetics—a scientific endeavor whose very core pivoted around the ontology of a machine.

In an attempt to nest the theoretical pillars of cybernetics within the evolution of ideas, the context that preceded the movement was laid out, starting with identifying an impetus for addressing certain fundamental mathematical problems presented at the beginning of the previous century by David Hilbert. Several European thinkers, ranging from Gottlob Frege to Kurt Gödel, engaged into an attempt of founding mathematics upon logical and/or formalistic grounds, triggering the foundational question for a rigorous definition of the notion of an algorithm. Alan Turing picked up the task, devising an abstract machine whose behavior would exactly define what an algorithm is supposed to do—and hence, what it is supposed to *be*. This definition of an entity, based upon its behavior, carried some far-reaching consequences. Turing's puzzling reversal regarding the possibility of intelligent machinery allows to show a possible link between cybernetics and the evolution of Turing's thought.

The chapters pertaining to cybernetics proper were intended to be, beyond a mere scholarly updated description of what the movement

© The Author(s) 2017
A. Malapi-Nelson, *The Nature of the Machine and the Collapse of Cybernetics*, Palgrave Studies in the Future of Humanity and its Successors, DOI 10.1007/978-3-319-54517-2_11

stood for, rather an articulated account of its theoretical pillars, based upon its different takes on the notion of a machine—namely, the latter being possibly teleological, possibly non-material and ultimately the instantiation of a theory. This reading of cybernetics' constitution and goals allowed me to show the subsequent internal tensions that arguably led to its demise. Coupled with the implosion of the Macy conferences and the Ratio club, two reported experiences occurring during the mid-1940s and lasting for less than a decade—those of William Ross Ashby and John von Neumann—can serve as historical markers pointing to the time when classical cybernetics was effectively over.

However, beyond historical clarifications, a theoretical issue of philosophical import seems to emerge from the scrutiny of both experiences. Both scientists, after struggling with physical instantiations of their advocated projects, ended up retreating from the material anchoring of these proposals into a disembodied realm—the very realm inhabited by those positions they were set to dispel in the first place, be it psychologisms or metaphysical truisms. These *ad intra* conflicting cybernetic developments are exposed and treated in the two chapters that followed, one pertaining to von Neumann's contributions and the other to Ross Ashby's—two cyberneticists whose influence on the enterprise has been so far relatively relegated to a secondary role, probably due in part to the widespread non-philosophical accounts of cybernetics' atrophy.

Both epic missions seemingly eventuated in negative outcomes. As advanced in this book, the tension between the mandate for model construction and the seemingly perennial temptation for abandoning all grips with physical reality might have proven to be devastating for the cybernetic endeavor. And more so if what ultimately characterized cybernetics was its impetus for control, in order to enhance and eventually complete its overarching articulating power. In the case of von Neumann, his kinematic model proved to be fraught with a plethora of practical issues pertaining to density, friction, heat, material fatigue, and so on. In a move to avoid the problematic entailments of the physical construction of a fairly complex entity, he resorted to leaving physical grounding aside, concentrating instead on the possible future logic that would allow such entities—his automata—to exist. In the case of Ashby, the acknowledgment of the need of augmentation of variables, for an entity—such as his homeostat—to more convincingly qualify as *alive* and *learning*, was realized in the construction of DAMS. However, one could witness via the dramatic entries in his diary how Ashby grew increasingly discouraged in the span of two years, observing a machine that only rendered chaotic behavior.

If one approaches the experience of classical cybernetics from the theoretical framework of an inquiry into the nature of a machine, one might be able to shed distinct light into what went down with such a peculiar enterprise—one that manages to carry allure until these days. The cybernetic articulation of a machine might have been ahead of its time, particularly in what regards the role of entropy in the physical world, and the relevance of materiality for a physical entity to exist. Martin Heidegger might have been somewhat prescient in his usually unattended preoccupation of cybernetics. He was aware that the technological mandate of reshaping manipulation, paradigmatically present in cybernetics, was purportedly pursued for the sake of the improvement of the human condition—as Wiener would proclaim in later years. However, as we saw, Heidegger recognized in cybernetics an epistemology that was going to wrap up, for him, the very history of metaphysics. Cybernetics was not only going to transcend modern science as it was practiced, bringing us one step closer to a Leibnitzian dream of a reality subsumed under the universal power of calculus. Cybernetics was also destined to replace philosophy itself, by means of the ultimate mechanization of anything understandable—precisely *because* it is frameable.

The last two chapters were devoted to philosophically explore the contemporary state of affairs in science and technology in what regards its possible influence in averting the fate experienced by classical cybernetics. A crucial aspect of this articulation was to unveil their transhumanistic character. Far from alienating or strange, transhumanism seems to be the necessary outcome of already familiar philosophies—now amenable of flourishing thanks to these technological developments, making the positive alteration of the human condition more attainable.

Whatever outcome may arrive regarding the status of "the human" from this deep restructuring of the technological impetus, it is useful to realize that the ideas driving these late scientific foresights and technological goals are not radically new—at least in spirit. Francis Bacon launched an effective mechanization of reality which shaped the way in which natural phenomena are scientifically dissected and systematized via an inter-operatively successful corpus of "explanatory mechanisms". For Giambattista Vico the ultimate signature for true knowledge becomes the capacity to literally build the object of knowledge, and so the total mechanization of explainable phenomena had to eventually—but necessarily—occur. Nature becomes understandable due to its mechanical essence. Moreover: A machine, by definition, can be tinkered with, fixed and improved. Thus, man can indeed be closer to its own potential. Indeed, as the Scotian univocity of language would suggest with reasonable stretch, divine qualities were always present in us to some degree. Now the angelic features dormant in us might begin to gradually—but finally—come to fruition. It would seem that the force transcending embodiment toward non-materiality is at last allowing for a purification of the best possible mold. After all, the intellectual tendency for disengagement from physical substrata toward an immaterial, "cleaner" and universal reality has been with us since Ancient Greece. Once that abstracted containment is filled up, it can be re-instantiated in physical reality—thus radically modifying it.

Thus, it is no wonder that agendas revolving around the Nano-Bio-Info-Cogno (NBIC) Convergence, exploiting the possibilities opened up by methodologies of bio-mechanization, virtual simulation and nano-construction, have high hopes for an upcoming Renaissance 2.0 of sorts. A scientifically viable instantiation of a collective hope for a positively transhumanist future seems to be in the making, regardless of whether the approach for its implementation takes a high-risk stance or one that advocates precaution. This dynamism of mechanization, purified abstraction and ulterior re-instantiation might indeed hold the key for our biological survival *vis-à-vis* a potential ecological or evolutionary catastrophe—even if in the process the traditional notion of "the human" gets trivialized into oblivion.

The awareness that "human nature"—whatever we understand by it—would indeed not survive the advance of science is a common sentiment among humanities scholars in the last decades. If we are reminded that the very forces behind these pervasive technological disruptions are deeply rooted into what prominent humanists proposed between the thirteenth and sixteenth centuries, the context for understanding this progress can be geared toward a more enriched and enhanced approach—even if what we obtain as an outcome is beyond canonically human recognition.

If we are to learn a lesson from what happened to classical cybernetics, putting out into the open the metaphysical tenets behind these "emergent technologies" could indeed further foster their advancement. Correspondingly, we can cultivate the awareness that there is a hopeful, transcendent, notion of what it entails to be human—which inspired these medieval and early modern intellectual forces in the first place.

The difference is that now we might be witnessing the beginning of the technological feasibility for this *next* human. If one is to thoroughly articulate the underlying humanistic impetus behind pervasively disruptive technologies, the input of the humanities could not only let itself be felt, but indeed guide the late scientific ethos toward biological alteration into—beyond survival—indeed human flourishing. When facing a new way of human existence, it will hopefully help us find a linearity with what makes us who we are. In fact, we will most likely have to put umprecedented effort in finding out what that is.

BIBLIOGRAPHY

Abraham, T. H. (2003). From theory to data: Representing neurons in the 1940s. *Biology and Philosophy, 18*(3), 415–426.

Académie Catholique de France (2015). *L'homme augmenté conduit-il au transhumanisme?*. Colloque annuel. Université Catholique de Lyon (Samedi 28 novembre).

Adkins, B. M. (1951). The homeostat. *British Journal for the Philosophy of Science, 2*(7), 248.

Alighieri, D. (1472). *The Divine Comedy: The Vision of Hell, Purgatory, and Paradise.* H. F. Cary (Trans.). Chicago, IL: Thompson & Thomas (1901 – originally published in 1814).

Ampère, A. M. (1838). *Essai sur la philosophie des sciences.* Paris, France: Bachelier.

Aquinas, T. (OP) (1273) *Summa Theologica.* Notre Dame, IN: Christian Classics (1981 – originally published in English in 1911).

Aristotle *De Generatione Animalium.* A. Platt (Trans.). Gloucestershire,UK: Clarendon Press. 1910.

Asaro, P. (2006). On the origins of the synthetic mind: Working models, mechanisms, and simulations. (Ph.D., University of Illinois at Urbana-Champaign).

Asaro, P. (2008). From mechanisms of adaptation to intelligence amplifiers: The philosophy of W. Ross ashby. In P. Husbands, O. Holland, & M. Wheeler (Eds.), *The Mechanical Mind in History* (pp. 149–184). Cambridge, MA: MIT Press.

Asaro, P. (2010). Whatever happened to cybernetics?. In G. Friesinger, J. Grenzfurthner, T. Ballhausen, & V. Bauer (Eds.), *Geist in der Maschine.*

© The Author(s) 2017   271
A. Malapi-Nelson, *The Nature of the Machine and the Collapse of Cybernetics*, Palgrave Studies in the Future of Humanity and its Successors, DOI 10.1007/978-3-319-54517-2

Medien, Prozesse und Räume der Kybernetik (pp. 39–49). Vienna, Austria: Verlag Turia & Kant.

Ashby, W. R. (1962). Principles of the self-organizing system. In H. Von Foerster & G. W. Zopf Jr. (Eds.), *Principles of Self-Organization: Transactions of the University of Illinois Symposium* (pp. 255–278). London, UK: Pergamon Press.

Ashby, W. R. (1928–1972). *Journal.* The W. Ross Ashby Digital Archive, 2008. Retrieved from www.rossashby.info/journal

Ashby, W. R. (1940). Adaptiveness and equilibrium. *Journal of Mental Science, 86,* 478–483.

Ashby, W. R. (1945). The physical origin of adaptation by trial and error. *Journal of General Psychology, 32,* 13–25.

Ashby, W. R. (1947a). The nervous system as physical machine: With special reference to the origin of adaptive behaviour. *Mind, 56*(1), 44–59.

Ashby, W. R. (1947b). Principles of the self-organizing dynamic system. *Journal of General Psychology, 37*(2), 125–128.

Ashby, W. R. (1949a). Experimental homeostat. *Electroencephalography & Clinical Neurophysiology, 1,* 116–117.

Ashby, W. R. (1949b, March). The Electronic Brain. *Radio-Electronics, XX*(6), 77–80.

Ashby, W. R. (1949c, January 24). The thinking machine. *Time LIII*(4), 66.

Ashby, W. R. (1952a). *Design for a Brain.* New York, NY: John Wiley & Sons.

Ashby, W. R. (1952b). Can a mechanical chess-player outplay its designer?. *British Journal for the Philosophy of Science, 3,* 44–57.

Ashby, W. R. (1953). Homeostasis. In H. Von Foerster, M. Mead, & H. L. Teuber (Eds.), *Cybernetics: Circular Causal and Feedback Mechanisms in Biological and Social Systems Transactions of the Ninth Conference, New York, March 20–21, 1952,* (pp. 73–108). New York, NY: Josiah Macy, Jr. Foundation.

Ashby, W. R. (1956). *An Introduction to Cybernetics.* London: UK: Chapman and Hall. New York, NY: John Wiley & Sons.

Ashby, W. R. (1960). *Design for a Brain: The Origin of Adaptive Behavior* (2nd revised ed.). New York, NY: John Wiley & Sons.

Ashby, W. R. (1981). *Mechanisms of Intelligence: Ross Ashby's Writings on Cybernetics* (R. Conant, Ed.). Salinas, CA: Intersystems Publications.

Aspray, W. (1990). The origins of John von Neumann's theory of automata. In J. Glimm, J. Impagliazzo, & I. Singer (Eds.). Proceedings of Symposia in Pure Mathematics, Volume 50 *The Legacy of John von Neumann,* Providence, RI: American Mathematical Society.

Bacon, F. (1603). Valerius terminus. In J. M. Robertson (Ed.), *The Philosophical Works of Francis Bacon* (pp. 186–205). London, UK: Routledge, 2011.

Bacon, F. (1620a). The new organon. In R. S. Sargent (Ed.), *Francis Bacon: Selected Philosophical Works* (pp. 86–189). Indianapolis, IN: Hackett, 1999.

Bacon, F. (1620b). The great instauration. In R. S. Sargent (Ed.), *Francis Bacon: Selected Philosophical Works* (pp. 66–85). Indianapolis, IN: Hackett, Vol. 1999,

Bell, E. T. (1937). *Men of Mathematics* (2nd ed.,Vol. 2). London, England: Penguin Books.

Bennett, S. (1993). *A History of Control Engineering, 1930–1955.* London, UK: Peter Peregrinus Inc.

Bennett, S. (1996). A brief history of automatic control. *IEEE Control Systems, 16*(3), 17–25.

Bernal, J. D. (1929). *The World, the Flesh & the Devil: An Enquiry into the Future of the Three Enemies of the Rational Soul.* Bloomington: Indiana University Press. 1969.

Berryman, S. (2003). Ancient Automata and Mechanical Explanation. *Phronesis, 48*(4), 344–369.

Berube, D. (2005). *Nano-Hype: The Truth Behind the Nanotechnology Buzz.* Amherst, NY: Prometheus Books.

Bissell, C. (2009). A history of automatic control. In S. Y. Nof (Ed.), *Springer Handbook of Automation* (pp. 53–69). Springer handbook series (LXXVI) Heidelberg, Germany: Springer Verlag.

Boden, M. A. (2006). *Mind as Machine: A History of Cognitive Science.* Oxford; New York: Clarendon Press; Oxford University Press.

Bostrom, N. (2005). A history of transhumanist thought. *Journal of Evolution and Technology, 14*(1), 1–25.

Bostrom, N. (2011). In defense of posthuman dignity. In G. R. Hansell & W. Grassie (Eds.), *H+/–: Transhumanism and its Critics* (pp. 55–66). Philadelphia, PA: Metanexus Institute.

Bowles, M. (1996). U.S. technological enthusiasm and British technological skepticism in the age of the analog brain. *IEEE Annals of the History of Computing, 18*(4), 5–15.

Brown, A. (2005). *J. D. Bernal: The Sage of Science.* Oxford, UK: Oxford University Press.

Bryant, J. (1991). *Systems Theory and Scientific Philosophy: An Application of the Cybernetics of W. Ross Ashby to Personal and Social Philosophy, The Philosophy of Mind, and the Problems of Artificial Intelligence.* Lanham, MD: University Press of America.

Burks, A. W. (1959). The Logic of Fixed and Growing Automata. *Proceedings of an International Symposium on the Theory of Switching,* 2–5 April *1957* (part I, pp. 147–188). Cambridge, MA: Harvard University Press.

Burks, A. W. (1960). Computation, behavior, and structure in fixed and growing automata. In M. Yovits & S. Cameron (Eds.), *Self-Organizing Systems* (pp. 282–311, 312–314). New York, NY: Pergamon Press.

Burks, A. W. (1963a). Programming and the theory of automata. In P. Braffort & D. Hirschberg (Eds.), *Computer Programming and Formal Systems* (pp. 100–117). Amsterdam, Holland: North Holland Publishing Company.

Burks, A. W. (1963b). Toward a theory of automata based on more realistic primitive elements. In C. M. Popplewell (Ed.), *Information Processing, Proceedings of IFIP Congress 62* (pp. 379–385). Amsterdam, Holland: North Holland Publishing Company.

Burks, A. W. (1966). Automata models of self-reproduction, *Information Processing. 7*(May, 1966), 121–123 (Information Processing Society of Japan).

Burks, A. (1969). *Von Neumann's self-reproducing automata: Technical report.* Ann Arbor, MI: University of Michigan, Office of Research Administration.

Burks, A. W. (Ed.). (1970). *Essays on Cellular Automata.* Urbana, IL: University of Illinois Press.

Burks, A. W. (1974). Cellular automata and natural systems. In W. D. Keidel, W. Handler, & M. Spreng (Eds.), *Cybernetics and Bionics Proceedings of the 5th Congress of the Deutsche Gesellschaft für Kybernetik, Nürnberg* March 28–30, *1973*, (pp. 190–204). Munich, Germany: R. Oldenbourg.

Burks, A. W. (1975). Logic, biology, and automata – Some historical reflections. *International Journal of Man-Machine Studies, 7*, 297–312.

Burks, A. W., & Wang, H. (1975). The logic of automata, *Journal of the Association for Computing Machinery. 4*(April, 1957), 193–218 and *4*(July, 1957), 279–297.

Catholic Church. (1999). *Catechism of the Catholic Church: Revised in Accordance with the Official Latin Text Promulgated by Pope John Paul II.* Vatican City: Libreria Editrice Vaticana.

Chapuis, A., & Gélis, E. (1928). *Le Monde des Automates: Étude Historique et Technique* (Vols. 1–2). Genève, Switzerland: Editions Slatkine.

Choi, B. C., & Pak, A. W. (2006). Multidisciplinarity, interdisciplinarity and transdisciplinarity in health research, services, education and policy: 1. Definitions, objectives, and evidence of effectiveness. *Clinical Investigative Medicine, 29*(6), 351–364.

Christensen, W. (1996). A complex systems theory of teleology. *Biology and Philosophy, 11*(3), 301–320.

Cole-Turner, R. (2011). *Transhumanism and Transcendence: Christian Hope in an Age of Technological Enhancement.* Washington, DC: Georgetown University Press.

Conway, F., & Siegelman, J. (2005). *Dark Hero of the Information Age: In Search f Norbert Wiener, the Father of Cybernetics.* New York, NY: Basic Books.

Copeland, J. (Ed.). (2004). *The Essential Turing: Seminal Writings in Computing, Logic, Philosophy, Artificial Intelligence, and Artificial Life plus The Secrets of Enigma.* New York, NY: Oxford University Press.

Copeland, J. (2005). Introduction. In J. Copeland et al. (Ed.), *Alan Turing's Electronic Brain: The Struggle to Build the ACE, the World's Fastest Computer* (pp. 1–14). Oxford, UK: Oxford University Press.

Copeland, J. (2012). *Turing: Pioneer of the Information Age*. Oxford, UK: Oxford University Press.

Copeland, J., & Proudfoot, D. (2005). Turing and the computer. In J. Copeland et al. (Ed.), *Alan Turing's Electronic Brain: The Struggle to Build the ACE, the World's Fastest Computer* (pp. 107–148). Oxford, UK: Oxford University Press.

Cordeschi, R. (2002). *The Discovery of the Artificial: Behavior, Mind, and Machines Before and Beyond Cybernetics*. Dordrecht; Boston: Kluwer.

Cordeschi, R. (2008a). Steps toward the synthetic method: Symbolic information processing and self-organizing systems in early artificial intelligence modeling. In P. Husbands, O. Holland, & M. Wheeler (Eds.), *The Mechanical Mind in History* (pp. 219–258). Cambridge, MA: MIT Press.

Cordeschi, R. (2008b). Cybernetics. In L. Floridi (Ed.), *The Blackwell Guide to the Philosophy of Computing and Information* (pp. 186–196). Oxford, UK: Blackwell.

Daly, T. T. W. (2011). Chasing Methuselah: Transhumanism and Christian *Theosis* in Critical Perspective. In R. Cole-Turner (Ed.), *Transhumanism and Transcendence: Christian Hope in an Age of Technological Enhancement* (pp. 131–144). Washington, DC: Georgetown University Press.

Daly, T. T. W. (2014). Diagnosing Death in the Transhumanism and Christian Traditions. In C. Mercer & T. Trothen (Eds.), *Religion and Transhumanism: The Unknown Future of Human Enhancement* (pp. 83–96). Santa Barbara, CA: ABC-CLIO.

Darwin, C. (1859). *On the Origin of Species by Means of Natural Selection, or the Preservation of Favoured Races in the Struggle for Life*. London, UK: John Murray.

Davis, M. (2000). *The Universal Computer: The Road from Leibniz to Turing*. New York, NY: W. W. Norton & Company Inc.

Davis, P., & Kenyon, D. H. (1989). *Of Pandas and People: The Central Question of Biological Origins*. 2nd ed. Dallas, TX: Haughton.

De Grey, A. (Ed.) (2004). Strategies for engineered negligible senescence: Why genuine control of aging may be foreseeable. *Annals of the New York Academy of Sciences, 1019*, xv–xvi, 1–592.

De Latil, P. (1956). *Thinking by Machine: A Study of Cybernetics*. Boston, MA: Houghton Mifflin.

Delio, I. (2012). Transhumanism or ultrahumanism? Teilhard de Chardin on technology, religion and evolution. *Theology and Science, 10*(2), 153–166.

Dennett, D. (1993). Review of *The Embodied Mind*, by F. Varela, E. Thompson and E. Rosch. *American Journal of Psychology, 106*, 121–126.

Descartes, R. (1649). Reply to More (February) (M. Grene & R. Ariew, Trans.). In R. Ariew (Ed.), *Rene Descartes: Philosophical Essays and Correspondence* (pp. 292–297). Indianapolis, IN: Hackett, 2000.

Descartes, R. (1662). Treatise on man (S. Gaukroger, Trans.). In S. Gaukroger (Ed.), *Rene Descartes: The World and Other Writings* (pp. 99–169). Cambridge, UK: Cambridge University Press, 1998.

Drexler, E. (1986). *Engines of Creation: The Coming Era of Nanotechnology*. New York, NY: Anchor Books.

Dronamraju, K. (1999). Erwin schrödinger and the origins of molecular biology. *Genetics, 153*(3), 1071–1076.

Dupuy, J. P. (2000). *The Mechanization of the Mind: On the Origins of Cognitive Science.* Princeton, N.J: Princeton University Press.

Dupuy, J. P. (2004). Do we shape technologies or do they shape us? *Proceedings of the Conference "Converging Technologies for a Diverse Europe"* (September 2004). Brussels, Belgium: European Commission.

Dupuy, J. P. (2007). Some pitfalls in the philosophical foundations of nanoethics. *Journal of Medicine and Philosophy, 32*(3), 237–261.

Dvorsky, G., & Hughes, J. (2008). *Postgenderism: Beyond the Gender Binary*. Hartford, CT: Institute for Ethics and Emerging Technologies.

Dyson, F. (2007). The future of biology. In F. Dyson (Ed.). *A Many-Colored Glass: Reflections on the Place of Life in the Universe* (ch. 1). Charlottesville, VA: University of Virginia Press.

Easterling, K. (2001). Walter pitts. *Cabinet, 5*(02), 33–36.

École polytechnique (France). Centre de recherche épistémologie et autonomie (1985). *Généalogies de l'auto-organisation.* Paris: Cahiers du C.R.E.A.

Edwards, P. N. (1996). The machine in the middle: Cybernetic psychology and World War II. In P.N. Edwards (Ed.). *The Closed World: Computers and the Politics of Discourse in Cold War America* (pp. 174–207). Cambridge, MA: MIT Press.

Eigler, D. M. & Schweizer, E. K. (1990). Positioning single atoms with a scanning tunneling microscope. *Nature* 344, 524–526.

Eisenstein, E. (2012). *The Printing Revolution in Early Modern Europe*. 2nd ed. New York, NY: Cambridge University Press.

ETC Group (2007). *Extreme Genetic Engineering: An Introduction to Synthetic Biology – Report* (Available at: etcgroup.org)

European Commission (2004a). *Converging Technologies: Shaping the Future of European Societies.* A. Nordmann (Rapporteur). Luxembourg: CORDIS.

European Commission (2004b). *Towards a European Strategy for Nanotechnology.* Luxembourg: CORDIS.

European Commission (2008). *Knowledge Politics and New Converging Technologies: A Social Science Perspective.* S. Fuller (Rapporteur). Luxembourg: CORDIS.

Ewald, W. B. (Ed.). (1996). *From Kant to Hilbert: A Source Book in the Foundations of Mathematics* (Vol. 2). New York, NY: Oxford University Press.

Findler, N. V. (1988). The debt of artificial-intelligence to John von Neumann. *Artificial Intelligence Review*, 2(4), 311–312.

Feynman, R. P. (1960). There's Plenty of Room at the Bottom. *Engineering and Science*, 23(5), 22–36.

Fisher, R. A., & Yates, F. (1963). *Statistical Tables for Biological, Agricultural and Medical Research.* London, UK: Oliver and Boyd. (Original work published in 1938.

Floridi, L. (2004). *The Blackwell Guide to the Philosophy of Computing and Information.* Malden, MA: Blackwell Pub.

Franchi, S., & Güzeldere, G. (2005). *Mechanical Bodies, Computational Minds: Artificial Intelligence from Automata to Cyborgs.* Cambridge, MA: MIT Press.

François, D. (2007). *The Self-Destruction of the West.* Paris, France: Éditions Publibook.

Freddolino, P. L., Arkhipov, A. S., Larson, S. B., McPherson, A., & Schulten, K. (2004). Molecular dynamics simulations of the complete satellite tobacco mosaic virus. *Structure, 14,* 437–449.

Frege, G. (1884). *The Foundations of Arithmetic: A Logico-Mathematical Enquiry Into the Concept of Number.* (J. L. Austin, Trans.). Evanston, IL: Northwestern University Press, 1980.

Freitas, R., & Gilbreath, W. (Eds.). (1980). Proceedings of the 1980 NASA/ASEE Summer Study: *Advanced Automation for Space Missions.* Washington, DC: US Government Printing Office.

Freitas, R. & Merkle, R. (2004). *Kinematic Self-Replicating Machines.* Georgetown, TX: Landes Bioscience.

Fuller, S. (2007). *Science Vs. Religion?: Intelligent Design and the Problem of Evolution.* Cambridge, UK: Polity.

Fuller, S. (2008). *Dissent Over Descent: Intelligent Design's Challenge to Darwinism.* Cambridge, UK: Icon.

Fuller, S. (2011). *Humanity 2.0: What it Means to be Human Past, Present and Future.* Basingstoke, UK: Palgrave Macmillan.

Fuller, S. (2013a). *Preparing for Life in Humanity 2.0.* Basingstoke, UK: Palgrave Macmillan.

Fuller, S. (2013b). Ninety degree revolution, *Aeon Magazine.* (24 October).

Fuller, S. (2016). Morphological Freedom and the Question of Responsibility and Representation in Transhumanism. *Confero, 4*(2), 33–45.

Fuller, S., & Lipinska, V. (2014). *The Proactionary Imperative: A Foundation for Transhumanism.* Basingstoke, UK: Palgrave Macmillan.

Galison, P. (1994). The ontology of the enemy: Norbert Wiener and the cybernetic vision. *Critical Inquiry, 21*(1), 228–266.

Galliott, J., & Lotz, M. (Eds.). (2016). *Super Soldiers: The Ethical, Legal and Social Implications.* Abingdon, UK: Routledge.

Gaukroger, S. (1995). *Descartes: An Intellectual Biography.* Oxford, UK: Oxford University Press.

Gerovitch, S. (2002). *From Newspeak to Cyberspeak: A History of Soviet Cybernetics.* Cambridge, MA: MIT Press.

Gibson, D. G., Benders, A. G., Andrews-Pfannkoch, C., Denisova, E. A., Baden-Tillson, H., Zaveri, J., Stockwell, T. B., Brownley, A., Thomas, D. W., Algire, M. A., Merryman, C., Young, L., Noskov, V. N., Glass, J. I., Venter, C. J., Hutchison, I. I. I. C. A., & Smith, H. O. (2008). Complete chemical synthesis, assembly, and cloning of a *Mycoplasma genitalium* genome. *Science, 319*(5867), 1215–1220.

Gibson, D. G., Glass, J. I., Lartigue, C., Noskov, V. N., Chuang, R.-Y., Algire, M. A., Benders, A. G., Montague, M. G., Ma, L., Moodie, M. M., Merryman, C., Vashee, S., Krishnakumar, R., Assad-Garcia, N., Andrews-Pfannkoch, C., Denisova, E. A., Young, L., Qil, Z.-Q., Segall-Shapiro, T. H., Calvey, C. H., Parmar, P. P., Hutchison, I. I. I. C. A., Smith, H. O., & Venter, C. J. (2010). Creation of a bacterial cell controlled by a chemically synthesized genome. *Science, 329*(5987), 52–56.

Glimm, J., Impagliazzo, J., & Singer, I. (Eds.). (1990). Proceedings of Symposia in Pure Mathematics, Volume 50: *The Legacy of John von Neumann.* Providence, RI: American Mathematical Society.

Gödel, K. (1931). On formally undecidable propositions of principia mathematica and related systems I (E. Mendelson, Trans.). In M. Davis (Ed.), (1965) *The Undecidable: Basic Papers on Undecidable Propositions, Unsolvable Problems and Computable Functions* (pp. 4–38). Hewlett, NY: Raven.

Gödel, K. (1934). On undecidable propositions of formal mathematical systems. In M. Davis (Ed.), (1965) *The Undecidable: Basic Papers on Undecidable Propositions, Unsolvable Problems and Computable Functions* (pp. 39–74). Hewlett, NY: Raven.

Goldstine, H. H. (1993). *The Computer from Pascal to von Neumann.* Princeton, NJ: Princeton University Press. (Original work published 1972).

Goujon, P. (1994). From logic to self-organization – The learning of complexity. *Cybernetica, 37*(3–4), 251–272.

Haldane, J. B. S. (1923). *Daedalus: Or, Science and the Future.* London, UK: Kegan, Paul, Trench, Trubner & Co., Ltd.

Haldane, J. S. (1922). *Respiration.* New Haven, CT: Yale University Press.

Harrison, P. (2002). Original sin and the problem of knowledge in early modern europe. *Journal of the History of Ideas, 63,* 239–259.

Harrison, P., & Wolyniak, J. (2015). The history of transhumanism. *Notes and Queries, 62*(3), 465–467.

Hawking, S. W. (Ed.). (2005). *God Created the Integers: The Mathematical Breakthroughs that Changed History.* Philadelphia, PA: Running Press.

Hayles, N. K. (1999). *How we Became Posthuman: Virtual Bodies in Cybernetics, Literature, and Informatics.* Chicago, IL: University of Chicago Press.

Hayward, R. (2001). The tortoise and the love-machine: Grey Walter and the politics of electroencephalography. *Science in Context, 14*(4), 615–641.

Heidegger, M. (1977a). The question concerning technology. In W. Lovitt (Trans.). *The Question Concerning Technology and Other Essays* (pp. 3–35). New York, NY: Harper & Row. (Original work published 1954).

Heidegger, M. (1977b). The end of philosophy and the task of thinking. In D. F. Krell (Ed.), *Basic Writings. From Being and Time (1927) to The Task of Thinking (1964)* (pp. 373–392). New York, NY: Harper & Row. (Original work published 1969).

Heidegger, M. (1981). "Only a God Can Save Us": The Spiegel interview. In T. Sheehan (Ed.), *Heidegger: The Man and the Thinker* (pp. 45–67). Chicago, IL: Precedent. (Original work published 1966).

Heidegger, M. (1984). *The Metaphysical Foundations of Logic* (M. Heim, Trans.). Bloomington, IN: Indiana University Press. (Original work published 1978).

Heidegger, M. (1999). *Contributions to Philosophy (From Enowning)* (P. Emad & K. Maly, Trans.). Bloomington, IN: Indiana University Press. (Original work published 1989).

Heidegger, M., & Fink, E. (1993) *Heraclitus Seminar* (C. Seibert, Trans.). Evanston, IL: Northwestern University Press. (Original work published 1970).

Heims, S. J. (1980). *John von Neumann and Norbert Wiener: From Mathematics to the Technologies of Life and Death.* Cambridge, MA: MIT Press.

Heims, S. J. (1991). *The Cybernetics Group.* Cambridge, MA: MIT Press.

Hertz, H. (1899). *The Principles of Mechanics Presented in a New Form* (D. E. Jones, J. T. Walley, & P. Lenard, Ed.). London, UK: Macmillan. (Original work published in 1894).

Heylighen, F. (2000). Foundations and Methodology for an Evolutionary World View: A Review of the Principia Cybernetica Project. *Foundations of Science, 5*(4), 457–490.

Hilbert, D. (1902). Mathematical problems. (M. W. Newson, Trans.). (Original work published in 1900), *Bulletin of the American Mathematical Society, 8*(10), 437–479.

Hodges, A. (1983). *Alan Turing: The Enigma of Intelligence.* London, UK: Unwin Publishing Company.

Hodges, A. (2000). Turing. In R. Monk & F. Raphael (Eds.), *The Great Philosophers: From Socrates to Turing* (pp. 493–540). London, U.K.: Phoenix.

Hodges, A. (2008). What did Alan Turing mean by "Machine"?. In P. Husbands, O. Holland, & M. Wheeler (Eds.), *The Mechanical Mind in History* (pp. 75–90). Cambridge, MA: MIT Press.

Holland, O., & Husbands, P. (2011). The origins of British cybernetics: The Ratio Club. *Kybernetes, 40*(1/2), 110–123.

Huchingson, J. E. (2005). *Religion and the Natural Sciences: The Range of Engagement.* Eugene, OR: Wipf and Stock (Original work published 1993).

Humes, E. (2007). *Monkey Girl: Evolution, Education, Religion, and the Battle for America's Soul*. New York, NY: Harper Collins.

Husbands, P., Holland, O., & Wheeler, M. (Eds.). (2008). *The Mechanical Mind in History*. Cambridge, MA: MIT Press.

Husbands, P., & Holland, O. (2008). The Ratio Club: A hub of British cybernetics. In P. Husbands, O. H., & M. Wheeler (Eds.), *The Mechanical Mind in History* (pp. 91–148). Cambridge, MA: MIT Press.

Husbands, P., & Holland, O. (2012). Warren McCulloch and the British cyberneticians. *Interdisciplinary Science Reviews, 37*(3), 237–253.

Husky, H. D. (2005). The ACE test assembly, the Pilot ACE, the Big ACE, and the Bendix G15. In J. Copeland et al., *Alan Turing's Electronic Brain: The Struggle to Build the ACE, the World's Fastest Computer* (pp. 281–296). Oxford, UK: Oxford University Press.

Hutchins, W. J. (2000). Warren Weaver and the launching of MT: Brief biographical note. In W. J. Hutchins (Ed.), *Early Years in Machine Translation: Memoirs and Biographies of Pioneers* (pp. 17–20). Amsterdam, Netherlands: John Benjamins.

Hutchison, I. I. I.,. C. A., Chuang, R.-Y., Noskov, V. N., Assad-Garcia, N., Deerinck, T. H., Ellisman, M. H., Gill, J., Kannan, K., Karas, B. J., Ma, L., Pelletier, J. F., Qi, Z.-Q., Richter, R. A., Strychalski, E. A., Sun, L., Suzuki, Y., Tsvetanova, B., Wise, K. S., Smith, H. O., Glass, J. I., Merryman, C., Gibson, D. G., & Venter, J. C. (2016). Design and synthesis of a minimal bacterial genome. *Science, 351*(6280), 1414.

Huxley, J. (1951). Knowledge, morality and destiny. *Psychiatry, 14,* 127–151.

Hyötyniemi, H. (2006, August). *Neocybernetics in Biological Systems*. Espoo, Finland: Helsinki University of Technology, Control Engineering Laboratory. Retrieved from www.control.hut.fi/rpt/r151isbn9789512286133.pdf

Institute for Advanced Study (Spring 2007) T. S. Eliot at the Institute for Advanced Study. *The Institute Letter*. Princeton, NJ.

Jackson, B. (1972). The greatest mathematician in the world: Norbert Wiener stories. *Western Folklore, 31*(1), 1–22.

Jeffress, L. A. (Ed.). (1951). *Cerebral Mechanisms in Behavior: The Hixon Symposium*. New York, NY: John Wiley & Sons.

Jennings, H. S. (1906). *Behavior of the Lower Organisms*. New York, NY: Columbia University Press.

Johnston, J. (2008). *The Allure of Machinic Life: Cybernetics, Artificial Life, and the New Ai*. Cambridge, MA: MIT Press.

Jung, R. (2005). Postmodern systems theory: A phase in the quest for a general system. In E. Kiss (Hg.). *Postmoderne und/oder Rationalität* (pp. 86–93). Kodolányi János Főiskola, Székesfehérvár.

Kant, I. (1787). *Critique of Pure Reason*. (P. Guyer & A. W. Wood, Trans. & Eds.). New York, NY: Cambridge University Press, 1998.

Karr, J. R., Sanghvi, J. C., Macklin, D. N., Gutschow, M. V., Jacobs, J. M., Bolival Jr., B., Assad-Garcia, N., Glass, J. I., & Covert, M. W. (2012). A whole-cell computational model predicts phenotype from genotype. *Cell, 150,* 389–401.

Kay, L. E. (2000). *Who Wrote the Book of Life?: A History of the Genetic Code.* Stanford, CA: Stanford University Press.

Kleene, S. C. (1967). *Mathematical Logic.* New York, NY: John Wiley & Sons.

Kurzweil, R. (2005). *The Singularity is Near: When Humans Transcend Biology.* New York, NY: Viking Penguin.

Lafferty, R. T. (1911).The philosophy of Dante. *Annual Reports of the Dante Society, 30,* 1–34.

Langdon, C. (1918). *The Divine Comedy of Dante Alighieri: The Italian Text with a Translation in English Blank Verse and Commentary.* Cambridge, MA: Harvard University Press.

Langton, C. (1989). Artificial Life. In C. Langton (Ed.), *Artificial Life* (pp. 1–44). Reading, MA: Addison-Wesley Publishing Company.

Lartigue, C., Glass, J. I., Alperovich, N., Pieper, R., Parmar, P. P., Hutchison, I. I. I. C. A., Smith, H. O., & Venter, J. C. (2007). Genome transplantation in bacteria: Changing one species to another. *Science, 317*(5838), 632–638.

Lassègue, J. (1996). What kind of Turing Test did Turing have in mind?. *Tekhnema, 3,* 37–58.

Latitude. (2012). *Robots @ School.* Ian Schulte (Rapporteur) (Available at latd.tv/ Latitude-Robots-at-School-Findings.pdf)

Lee, J. A. (2000). *The Scientific Endeavor: A Primer on Scientific Principles and Practice.* San Francisco: Addison Wesley Longman, Inc.

Lenski, R. E., Ofria, C., Pennock, R. T., & Adami, C. (2003). The evolutionary origin of complex features. *Nature, 423,* 134–144.

Leo XIII, P. (1879). *Aeterni Patris: Encyclical on the Restoration of Christian Philosophy.* Vatican City: Libreria Editrice Vaticana.

MacColl, L. A. (1945). *Fundamental Theory of Servomechanisms.* New York, NY: D. Van Nostrand.

MacDonald, I. (2008). *Adorno and Heidegger: Philosophical Questions.* Stanford, CA: Stanford University Press.

Macrae, N. (1992). *John von Neumann: The Scientific Genius who Pioneered the Modern Computer, Game Theory, Nuclear Deterrence, and Much More.* Providence, RI: American Mathematical Society.

Mahfood, S. (O. P.). (2015). *The Summa in Verse. Dante's Divine Comedy: Thomistic Philosophy in Narrative.* (Available at summainverse.wordpress. com/2015/03/26/the-summa-in-verse).

Malapi-Nelson, A. (2014). Humanities' metaphysical underpinnings of late frontier scientific research. *Humanities, 3*(4), 740–765.

Mancosu, P. (Ed.). (1998). *From Brouwer to Hilbert: The Debate on the Foundations of Mathematics in the 1920s.* New York, NY: Oxford University Press.

Mann, R. W. (1997). Sensory and motor prostheses in the aftermath of wiener. In V. Mandrekar & P. R. Masani (Eds.), *Proceedings of Symposia in Applied Mathematics Volume 52: Proceedings of the Norbert Wiener Centenary Congress, 1994: Michigan State University, November 27-December 3, 1994* (pp. 401–426). Providence, RI: American Mathematical Society.

Marciszewski, W., & Murawski, R. (1995). *Mechanization of Reasoning in a Historical Perspective.* Amsterdam, Netherlands; Atlanta, GA: Rodopi.

Marcovich, M. (1967). *Heraclitus. Greek Text with a Short Commentary. Editio maior.* Merida, Venezuela: Universidad de los Andes Press.

Margolis, E., & Laurence, S. (Eds.). (2007). *Creations of the Mind: Theories of Artifacts and Their Representation.* Oxford, UK: Oxford University Press.

Marsalli, M. (2006). McCulloch-Pitts neurons (National Science Foundation Grants #9981217 and #0127561). Retrieved from www.mind.ilstu.edu/curriculum/modOverview.php?modGUI=212

Masani, P. R. (1990). *Norbert Wiener, 1894–1964.* Basel, Switzerland: Birkhäuser.

Masani, P. R. (1994). The scientific methodology in the light of cybernetics. *Kybernetes, 23*(4), 5–132.

Mathews, S. (2008). *Theology and Science in the Thought of Francis Bacon.* Hampshire, UK: Ashgate.

Maxwell, J. C. (1868). On Governors. *Proceedings of the Royal Society of London, 16,* 270–283.

Mayr, O. (1970). *The Origins of Feedback Control.* Cambridge, MA: MIT Press.

McCarthy, J., Minsky, M. L., & Shannon, C. E. (1955). A proposal for the Dartmouth summer research project on artificial intelligence – August 31, 1955. In R. Chrisley & S. Begeer(Eds.). (2000). *Artificial intelligence: Critical concepts* (Vol. 2, pp. 44–53). London, UK & New York, NY: Routledge.

McCorduck, P. (2004). *Machines Who Think: A Personal Inquiry into the History and Prospects of Artificial Intelligence.* Natick, MA: A K Peters.

McCulloch, W. S. (1947). *An Account of the First Three Conferences on Teleological Mechanisms.* New York, NY: Josiah Macy, Jr. Foundation.

McCulloch, W. S. (1951). Why the mind is in the head [and discussion]. In L. A. Jeffress (Ed.), *Cerebral Mechanisms in Behavior: The Hixon Symposium* (pp. 42–74). New York, NY: John Wiley & Sons.

McCulloch, W. S. (1960). *General Semantics Bulletin, 26/27,* 7–18.

McCulloch, W. S. (1965). *Embodiments of Mind.* Cambridge, MA: MIT Press.

McCulloch, W. S. (1974). Recollections of the many sources of cybernetics. *ASC Forum, VI*(2), 5–16.

McCulloch, W. S., & Pitts, W. (1943). A logical calculus of the ideas immanent in nervous activity. *Bulletin of Mathematical Biophysics, 5,* 115–133.

McCulloch, W. S., & Pitts, W. (1948). The statistical organization of nervous activity. *Biometrics, 4,* 91–99.

Medina, E. (2006). Designing freedom, regulating a nation: Socialist cybernetics in Allende's Chile. *Journal of Latin American Studies*, *38*(3), 571–606.

Medina, E. (2011). *Cybernetic Revolutionaries: Technology and Politics in Allende's Chile*. Cambridge, MA: MIT Press.

Mercer, C., & Trothen, T. (Eds.). (2014). *Religion and Transhumanism: The Unknown Future of Human Enhancement*. Santa Barbara, CA: ABC-CLIO.

Mindell, D. (1995). Automation's finest hour: Bell laboratories' control systems in World War II. *IEEE Control Systems*, *15*(6), 72–80.

Mindell, D. (2002). *Between Human and Machine: Feedback, Control, and Computing Before Cybernetics*. Baltimore, MD: Johns Hopkins University Press.

Mindell, D., Segal, J., & Gerovitch, S. (2003). Cybernetics and information theory in the United States, France and the Soviet Union. In M. Walker (Ed.), *Science and Ideology: A Comparative History* (pp. 66–95). London, UK: Routledge.

Miner, R. (1998). Verum-factum and practical wisdom in the early writings of Giambattista Vico. *Journal of the History of Ideas*, *59*, 53–73.

Miner, R. (2004). *Truth in the Making: Creative Knowledge in Theology and Philosophy*. London UK: Routledge.

Mirowski, P. (2002). *Machine Dreams: Economics Becomes a Cyborg Science*. Cambridge, UK: Cambridge University Press.

More, M. (2013a). The philosophy of transhumanism. In M. More & N. Vita-More (Eds.), *The Transhumanist Reader: Classical and Contemporary Essays on the Science, Technology, and Philosophy of the Human Future* (pp. 3–17). West Sussex: John Wiley & Sons.

More, M. (2013b). The proactionary principle. In M. More & N. Vita-More (Eds.), *The Transhumanist Reader: Classical and Contemporary Essays on the Science, Technology, and Philosophy of the Human Future* (pp. 258–267). West Sussex: John Wiley & Sons.

Moschovakis, Y. N. (2001). What is an algorithm?. In B. Engquist & W. Schmid (Eds.), *Mathematics Unlimited – 2001 and Beyond* (pp. 919–936). Berlin, Germany: Springer.

Müller, A., & Müller, K. (Eds.). (2007). *An Unfinished Revolution? Heinz von Foerster and the Biological Computer Laboratory (BCL), 1958–1976*. Vienna, Austria: Edition Echoraum.

Namba, K. (2010). Conformational change of flagellin for polymorphic super-coiling of the flagellar filament. *Nature Structural & Molecular Biology*, *17*, 417–422.

Namba, K. (May 2002). Self-assembly of bacterial flagella. Paper presented at the *Annual Meeting of the American Crystallographic Association*, San Antonio, TX, USA.

National Intelligence Council (2012). *Global Trends 2030: Alternative Worlds*. Washington, DC: CreateSpace Independent Publishing Platform.

National Nanotechnology Initiative (2014a). *Supplement to the President's 2015 Budget.*Arlington, VA: National Nanotechnology Coordination Office.

National Nanotechnology Initiative (2014b). *Strategic Plan.* Arlington, VA: National Nanotechnology Coordination Office.

National Nanotechnology Initiative (Nano.org). *Nanotechnology 101: What is so Special About Nanotechnology?* (Available at: www.nano.gov/nanotech-101/special)

National Science Foundation (2003). *Converging Technologies for Improving Human Performance: Nanotechnology, Biotechnology, Information Technology and Cognitive Science.* W. S. Bainbridge & M. C. Roco (Eds.). Dordrecht: Kluwer.

Newton, I. (1687). *The Principia: Mathematical Principles of Natural Philosophy.* (B. Cohen, A. Whitman, & J. Budenz, Trans.). Oakland, CA: University of California Press, 1999.

Nobelprize.org. (2014a) "The Nobel Prize in Physics 1981". Nobel Media AB 2014. (Available at: www.nobelprize.org/nobel_prizes/physics/laureates/1986)

Nobelprize.org. (2014b) "The Nobel Prize in Chemistry 2016". Nobel Media AB 2014. (Available at: www.nobelprize.org/nobel_prizes/chemistry/laureates/2016)

Noble, A. (1997). *The Religion of Technology: The Divinity of Man and the Spirit of Invention.* New York, NY: Alfred A. Knopf, Inc.

Owens, L. (1986). Vannevar Bush and the differential analyzer: The text and context of an early computer. *Technology and Culture, 27*(1), 63–95.

Paley, W. (1802). *Natural Theology: Or, Evidences of the Existence and Attributes of the Deity, Collected from the Appearances of Nature.* Oxford, UK: Oxford University Press. 2006.

Pennock, R. (2007). Models, simulations, instantiations, and evidence: The case of digital evolution. *Journal of Experimental & Theoretical Artificial Intelligence, 19,* 29–42.

Piccinini, G. (2002). Review of the mechanization of mind: On the origins of cognitive science. by J. P. Dupuy, *Minds and Machines, 12*(3), 449–453.

Piccinini, G. (2004). The first computational theory of mind and brain: A close look at McCulloch and Pitts' "Logical calculus of ideas immanent in nervous activity". *Synthese, 141*(2), 175–215.

Pickering, A. (2002). Cybernetics and the mangle: Ashby, Beer and Pask. *Social Studies of Science, 32*(3), 413–437.

Pickering, A. (2005). A gallery of monsters: Cybernetics and self-organisation, 1940–1970. In S. Franchi & G. Güzeldere (Eds.), *Mechanical Bodies, Computational Minds: Artificial Intelligence from Automata to Cyborgs* (pp. 229–245). Cambridge, MA: MIT Press.

Pickering, A. (2010). *The Cybernetic Brain: Sketches of Another Future.* Chicago, IL: University of Chicago Press.

Polanyi, M. (1952). The hypothesis of cybernetics. *British Journal for the Philosophy of Science, 2,* 312–314.

Popper, K. (2005). *Unended Quest: An Intellectual Autobiography.* Oxfordshire, UK: Taylor and Francis (First published in 1974).

Poundstone, W. (1985). *The Recursive Universe: Cosmic Complexity and the Limits of Scientific Knowledge.* 1st ed. New York, NY: William Morrow & Co.

Prigogine, I., & Stengers, I. (1984). *Order Out of Chaos: Man's New Dialogue with Nature.* New York, NY: Bantam Books.

Ranisch, R., & Sorgner, S. L. (Eds.). (2014). *Post- and Transhumanism: An Introduction.* Frankfurt, Germany: Peter Lang GmbH.

Read, D. (2009). *Artifact Classification: A Conceptual and Methodological Approach.* Walnut Creek, CA: Left Coast Press.

Rédei, M. (2005). *John von Neumann: Selected Letters.* Introductory comments. In J. Von Neumann & M. Rédei (Ed.), (pp. 1–40). Providence, R.I: American Mathematical Society.

Rolston III, H. (2006). *Science and Religion: A Critical Survey.* West Conshohocken, PA: Templeton Foundation. (Original work published 1987).

Rosenblatt, F. (1958). The perceptron: A probabilistic model for information storage and organization in the brain. *Psychological Review, 65*(6), 386–408.

Rosenblueth, A., Wiener, N., & Bigelow, J. (1943). Behavior, purpose and teleology. *Philosophy of Science, 10,* 18–24.

Rosenblueth, A., & Wiener, N. (1945). The role of models in science. *Philosophy of Science, 12,* 316–321.

Rosenblueth, A., & Wiener, N. (1950). Purposeful and non-purposeful behavior. *Philosophy of Science, 17,* 318–318.

Royal Academy of Engineering (*Nanoscience and Nanotechnologies: Opportunities and Uncertainties*). London, UK: The Royal Society. 2004.

Sandberg, A. (2013). Morphological freedom – Why we not just want it, but need it. In M. More & N. Vita-More (Eds.), *The Transhumanist Reader: Classical and Contemporary Essays on the Science, Technology, and Philosophy of the Human Future* (pp. 56–64). West Sussex: John Wiley & Sons.

Schlatter, M., & Aizawa, K. (2008). Walter Pitts and "A logical calculus". *Synthese, 162*(2), 235–250.

Schrodinger, E. (2012). *What is Life?: With Mind and Matter and Autobiographical Sketches.* Cambridge, UK: Cambridge University Press. (Original work published 1944).

Scotus, D. (1300). *Philosophical Writings: A Selection.* Indianapolis, IN: Hackett. 1987.

Shanker, S. G. (1987a). *The Decline and Fall of the Mechanist Metaphor.* New York: St: Martin's.

Shanker, S. G. (1987b). *Wittgenstein and the Turning-Point in the Philosophy of Mathematics.* Albany, NY: SUNY Press.

Shanker, S. G. (1987c). Wittgenstein versus Turing on the nature of Church's thesis. *Notre Dame Journal of Formal Logic, 28,* 615–649.

Shanker, S. G. (1988). The dawning of (machine) intelligence. *Philosophica (Belgium), 42,* 93–144.

Shanker, S. G. (1995). Turing and the origins of AI. *Philosophia Mathematica, 3* (1), 52–85.

Shannon, C. E. (1948). A mathematical theory of communication. *The Bell System Technical Journal, 27,* 379–423.

Shannon, C. E. (1953). Computers and automata. *Proceedings of the IRE, 41*(10), 1234–1241.

Shannon, C. E., McCarthy, J., & Ashby, W. R. (1956). *Automata Studies.* Princeton, NJ: Princeton University Press.

Shannon, C. E. (1958). Von Neumann's contributions to automata theory. *Bulletin of the American Mathematical Society, 64*(3), 123–129.

Shannon, C. E., & Weaver, W. (1949). *The Mathematical Theory of Communication.* Champaign, IL: University of Illinois Press.

Shaw, R. E., & McIntyre, M. (1974). Algoristic foundations to cognitive psychology. In W. B. Weimer & D. S. Palermo (Eds.), *Cognition and the Symbolic Processes* (pp. 305–362). Hillsdale, NJ: Lawrence Erlbaum Associates.

Shirai, Y., Osgood, A. J., Zhao, Y., Kelly, K. F., & Tour, J. M. (2005). Directional control in thermally driven single-molecule nanocars. *Nano Letters, 5*(11), 2330–2334.

Sholl, D. A. (1956). *The Organization of the Cerebral Cortex.* New York, NY: John Wiley & Sons.

Simon, H. A. (1954). The axiomatization of classical mechanics. *Philosophy of Science, 21,* 340–343.

Simon, H. A. (1981). *The Sciences of the Artificial.* 2nd ed. Cambridge, MA: MIT Press.

Smalheiser, N. R. (2000). Walter Pitts. In. *Perspectives in Biology and Medicine, 43*(2), 217–226.

Smith, B. C. (2002). The foundations of computing. In M. Scheutz (Ed.), *Computationalism: New Directions.* Cambridge, MA: MIT Press.

Smith, P. (2007). *An Introduction to Gödel's Theorems.* Cambridge, UK: Cambridge University Press.

Sonleitner, F. (1986). What did Karl Popper really say about Evolution? *Creation Evolution, 6*(2), 9–14.

Sorgner, S. L. (2009). Nietzsche, the Overhuman, and Transhumanism. *Journal of Evolution and Technology, 20*(1), 29–42.

Spencer, C. (2009, 11 September). Profile: Alan Turing. *BBC News* (Available at news.bbc.co.uk/1/hi/uk/8250592.stm)

StatNano. (2016). *Annual Report 2015* (Available at statnano.com/publications/ 3864)

Steinhart, E. (2008). Teilhard de Chardin and Transhumanism. *Journal of Evolution and Technology*, 20(1), 1–22.

Stoll, R. R. (1979). *Set Theory and Logic*. Mineola, NY: Dover.

Tamburrini, G., & Datteri, E. (2005). Machine experiments and theoretical modelling: From cybernetic methodology to neuro-robotics. *Minds and Machines*, 15(3–4), 335–335.

Taylor, R. (1950a). Comments on a mechanistic conception of purposefulness. *Philosophy of Science*, 17(4), 310–317.

Taylor, R. (1950b). Purposeful and non-purposeful behavior: A rejoinder. *Philosophy of Science*, 17(4), 327–332.

Tedre, M. (2006). *The Development of Computer Science: A Sociocultural Perspective*. (Unpublished doctoral dissertation) University of Joensuu, Finland.

Tirosh-Samuelson, H. (2011). Engaging transhumanism. In G. R. Hansell & W. Grassie (Eds.), *H+/−: Transhumanism and Its Critics* (pp. 19–52). Philadelphia, PA: Metanexus Institute.

Turing, A. M. (1937). On Computable Numbers, with an Application to the Entscheidungsproblem. *Proceedings of the London Mathematical Society*, 42(1), 230–265.

Turing, A. M. (1938). On Computable Numbers, with an Application to the Entscheidungsproblem – A Correction. *Proceedings of the London Mathematical Society*, 43(1), 544–546.

Turing, A. M. (1939). Systems of logic based on ordinals. *Proceedings of the London Mathematical Society*, 45(1), 161–228.

Turing, A. M. (1946). Proposed electronic calculator. [paper, 48 sh. in envelope]. *The Papers of Alan Mathison Turing* (ref.: DSIR 10/385). King's College Archive Centre, Cambridge, U.K. (Xerox TS available at turingarchive.org; ref.: AMT/C/32)

Turing, A. M. (1947). Lecture to the London mathematical society on 20 February 1947. In S. B. Cooper & J. Van Leeuwen (Eds.), (2013) *Alan Turing: His Work and Impact* (pp. 486–497). Amsterdam, Netherlands: Elsevier.

Turing, A. M. (1948). Intelligent machinery. In J. Copeland (Ed.), (2004) *The Essential Turing: Seminal Writings in* Computing, Logic, Philosophy, Artificial Intelligence, and *Artificial Life plus The Secrets of Enigma* (pp. 395–432). New York, NY: Oxford University Press.

Turing, A. M. (1950). Computing machinery and intelligence. *Mind*, 59, 433–460.

Turing, A. M. (1996). Intelligent machinery, A heretical theory. *Philosophia Mathematica*, 3(4), 256–260.

UNESCO (1950). *The Race Question; UNESCO and its Programme*. (Vol. 3). Available at. unesdoc.unesco.org/images/0012/001282/128291eo.pdf.

United States Office of Naval Research. (1962). Transactions of the University of Illinois Symposium on Self-Organization: *Principles of self-organization*. H. Von Foerster & G. W. Zopf (Eds.). New York, NY: Pergamon Press.

United States Patent and Trademark Office. (2014, March). *Manual of Patent Examining Procedure*. Alexandria, VA: U.S. Government Printing Office.

Varela, F., Thompson, E., & Rosch, E. (1991). *The Embodied Mind: Cognitive Science and Human Experience.*. Cambridge, MA: MIT Press.

Vico, G. (1710). *On the Most Ancient Wisdom of the Italians: Unearthed from the Origins of the Latin Language*. L. M. Palmer (Trans.). Ithaca, NY: Cornell University Press, 1988.

Vico, G. (1731). *The Autobiography of Giambattista Vico*. M. H. Fisch & T. G. Bergin (Trans.). Ithaca, NY: Cornell University Press, 1963.

Vico, G. (1744). *The New Science of Giambattista*. *Vico*. T. G. Bergin & M. H. Fisch (Trans.). Ithaca, NY: Cornell University Press, 1988.

Von Bertalanffy, L. (1968). *General System theory: Foundations, Development, Applications*. New York, NY: George Braziller.

Von Foerster, H., Mead, M., & Teuber, H. L. (Eds.). (1950). *Cybernetics: Circular Causal and Feedback Mechanisms in Biological and* Social Systems.. Transactions of the Sixth Conference, New York, March 24–25, *1949*. New York, NY: Josiah Macy, Jr. Foundation.

Von Foerster, H., Mead, M., & Teuber, H. L. (Eds.). (1951). *Cybernetics: Circular Causal and Feedback Mechanisms in Biological and* Social Systems. Transactions of the Seventh Conference, New York, March 23–24, *1950*. New York, NY: Josiah Macy, Jr. Foundation.

Von Foerster, H., Mead, M., & Teuber, H. L. (Eds.). (1952). *Cybernetics: Circular Causal and Feedback Mechanisms in Biological and* Social Systems. Transactions of the Eighth Conference, New York, March 15–16 *1951*. New York, NY: Josiah Macy, Jr. Foundation.

Von Foerster, H., Mead, M., & Teuber, H. L. (Eds.). (1953). *Cybernetics: Circular Causal and Feedback Mechanisms in Biological and* Social Systems. Transactions of the Ninth *Conference, New York, March 20–21, 1952*. New York, NY: Josiah Macy, Jr. Foundation.

Von Foerster, H., Mead, M., & Teuber, H. L. (Eds.). (1955). *Cybernetics: Circular Causal and Feedback Mechanisms in Biological and* Social Systems. Transactions of the Tenth *Conference, Princeton, April 22, 23 and 24, 1953*. New York, NY: Josiah Macy, Jr. Foundation.

Von Foerster, H. (2003). *Understanding Understanding: Essays on Cybernetics and Cognition*. New York: Springer.

Von Neumann, J. (1945). *First Draft of a Report on the EDVAC*. (Contract No. W-670-ORD-4926, Between the United States Army Ordinance Department and the University of Pennsylvania). Philadelphia, PA: Moore School of Electrical Engineering.

Von Neumann, J. (1951). The general and logical theory of automata. In L. A. Jeffress (Ed.), *Cerebral Mechanisms in Behavior: The Hixon Symposium* (pp. 1–41). New York, NY: John Wiley & Sons.

Von Neumann, J. (1961). *John von Neumann. Collected works* (A. H. Taub, Ed.). New York, NY: Pergamon Press.

Von Neumann, J. (1966). *Theory of Self-Reproducing Automata* (A. W. Burks, Ed.). Urbana, IL: University of Illinois Press.

Von Neumann, J., Aspray, W., & Burks, A. W. (Eds.). (1987). *Papers of John von Neumann on computing and computer theory.* Cambridge, MA; Los Angeles, CA: MIT Press; Tomash Publishers.

Von Neumann, J. (2005). Letter to wiener. In M. Rédei (Ed.), *John von Neumann: Selected letters* (pp. 277–282). Providence, R.I: American Mathematical Society.

Wallerstein, I. (2003). America and the World: The Twin Towers as Metaphor. In I. Wallerstein *Decline of American Power: The U.S. in a Chaotic World* (ch. 9). New York, NY; London UK: The New Press.

Waters, B. (2011). Whose Salvation? Which Eschatology? Transhumanism and Christianity as Contending Salvific Religions. In R. Cole-Turner (Ed.), *Transhumanism and Transcendence: Christian Hope in an Age of Technological Enhancement* (pp. 163–176). Washington, DC: Georgetown University Press.

Weaver, W. (1949). The mathematics of communication. *Scientific American, 181* (1), 11–15.

Weaver, W. (1970). Molecular biology: Origin of the term. *Science, 170*, 581–582.

Wheeler, M. (2008). God's machines: Descartes on the mechanization of mind. In P. Husbands, O. Holland, & M. Wheeler (Eds.), *The Mechanical Mind in History* (pp. 307–330). Cambridge, MA: MIT Press.

Whitehead, A. N., & Russell, B. (1927). *Principia Mathematica, 3 Volume Set.* 2nd ed. Cambridge, UK: Cambridge University Press.

Wiener, N. (1948a). *Cybernetics: Or Control and Communication in the Animal and the Machine.* Paris, France: Hermann et Cie. Cambridge, MA: MIT Press. (2nd rev. ed. 1961).

Wiener, N. (1948b). Time, communication and the nervous system. In *Annals of the New York Academy of Sciences, 50*, 197–220.

Wiener, N. (1949). *The Extrapolation, Interpolation and Smoothing of Stationary Time Series: With Engineering Applications.* Cambridge, MA: MIT Press.

Wiener, N. (1950). *The Human Use of Human Being: Cybernetics and Society.* Boston, MA: Houghton Mifflin Co. (2nd rev. ed. 1954.

Wiener, N. (1956). *I Am Mathematician: The Later Life of a Prodigy.* New York, NY: Doubleday.

Wiener, N. (1964). *God and Golem, Inc: A Comment on Certain Points where Cybernetics Impinges on Religion.* Cambridge, MA: MIT Press.

Williams, J. D. (2013). "It's just a theory": Trainee science teachers' misunderstandings of key scientific terminology. *Evolution: Education and Outreach, 6*(12), 1–9.

Wisdom, J. O. (1951). The hypothesis of cybernetics. *British Journal for the Philosophy of Science, 2*, 1–24.

Wittgenstein, L. (1980). *Remarks on the Philosophy of Psychology* (Vol. 1). (G. E. M. Anscombe Trans., Ed. & G. H. von Wright, Ed.). Chicago, IL: University of Chicago Press.

Woese, C. R. (2004). A new biology for a new century. *Microbiology and Molecular Biology Reviews, 68*(2), 173–186.

Woese, C. R. (2007). Biology's next revolution. *Nature, 445*, 369.

Young, G. (2012). *The Russian Cosmists: The Esoteric Futurism of Nikolai Fedorov and His Followers*. New York, NY: Oxford University Press.

INDEX

κυβερνήτης, 48, 49

A

AA-Predictor, 9–24
Aberdeen Proving Ground, 31, 131, 136
Adaptiveness, 142, 145, 147, 211
Algorithm, 89–91, 94, 97, 98, 102, 104, 107, 118, 136, 140, 196n78, 208, 265
Alighieri, Dante, 231
Ampère, André-Marie, 48, 49n3
Analytical propositions, 82
Angels, 94, 206
Anthropomorphic, 111
Anti-aircraft, 5–9, 12, 14, 15, 25, 31, 34, 36–38, 63n49, 66n54, 122, 123, 136, 137, 208
Aquinas, Saint Thomas, 57n30, 229n46, 231n53, 237, 238, 239
Arithmetic, 32, 50n6, 81, 83, 84–85, 87, 88, 102
Artificial dichotomy, 1
Artificial Intelligence, 64, 67, 68n58, 98, 100, 106, 108, 109, 111n68, 133n76, 196n78, 219, 219n12, 233n60, 250

Ashby, William Ross, 2, 3, 19n25, 44, 57n29, 58, 65n53, 80, 121, 129, 135, 138, 139–168, 181, 187, 198, 203–207, 209–212, 214, 216, 227, 254, 257, 260, 262, 263, 266
Automata, 41, 96, 184, 184–185n42, 192, 192n66, 195n77, 196, 198–200, 207, 216, 266
Automaton, 95, 140, 195, 196, 197n80, 198, 200n88, 212, 216
Axiomatizing, 187, 197

B

Bacon, Francis, 94, 133n77, 240–244, 247, 253, 267
Baconian, 218n7, 240–243, 247, 253, 258
Bacterial flagellum, 254, 255, 255n20, 256n26, 259
Bainbridge, William, 223
Barbara, 72, 74, 75
Bateson, Gregory, 26, 36, 42, 61, 162
Beekman Hotel, 33, 56n28

© The Author(s) 2017
A. Malapi-Nelson, *The Nature of the Machine and the Collapse of Cybernetics*, Palgrave Studies in the Future of Humanity and its Successors, DOI 10.1007/978-3-319-54517-2

The manufacturer's authorised representative in the EU is Springer
Nature Customer Service Centre GmbH, Europaplatz 3, 69115 Heidelberg,
Germany. If you have any concerns regarding our products, please
contact ProductSafety@springernature.com

Printed and bound by CPI Group (UK) Ltd, Croydon, CR0 4YY
27/04/2026
02097624-0002